AF410857

Five Insights for Avoiding Global Collapse

Five Insights for Avoiding Global Collapse

What a 50-Year-Old Model of the World Taught Me About a Way Forward for Us Today

Gaya Herrington

Basel • Beijing • Wuhan • Barcelona • Belgrade • Novi Sad • Cluj • Manchester

Author
Gaya Herrington
Vice-President ESG Research,
Schneider Electric
Washington DC, USA

Editorial Office
MDPI
St. Alban-Anlage 66
4052 Basel, Switzerland

For citation purposes, cite as indicated below:

Herrington, Gaya. 2022. *Five Insights for Avoiding Global Collapse: What a 50-Year-Old Model of the World Taught Me About a Way Forward for Us Today.* Basel: MDPI, Page Range.

ISBN 978-3-0365-3722-1 (Hbk)
ISBN 978-3-0365-3721-4 (PDF)
doi: 10.3390/books978-3-0365-3721-4

Dedication

To environmental and social activists around the world. I have been thanked for my courage and persistence in my work on *The Limits to Growth*. I am grateful for these praises, but I do not deserve them. Although research and writing are excruciating and always thrice the amount of effort I thought they would be, doing so never made me feel unsafe. Environmental and human rights activists make enormous sacrifices protecting those with little or no voice. Each week, an average of four activists pay for this with their life, a third of which are indigenous people. The number of murders has been increasing for almost two decades. This is a humble salute to the bravery of all these environmental and social defenders. You are the unsung heroes of our time.

Gaya Herrington

Foreword

It is my pleasure to congratulate Gaya Herrington on her excellent research. Since last year, Gaya's work has often been characterized in the media as a validation of *The Limits to Growth*. More than anything, I personally find her research a testimony of what happens when scientists let their curiosity guide their work. In this situation, many find meaning in servitude to the world, like when my friends and I wrote *The Limits to Growth*.

Over the past five decades, I have witnessed the waves of debate that *The Limits to Growth* has stirred. The voices of support and the choirs of criticism. Gaya has brought some order to this debate by doing the only thing that is scientifically rigorous when one wants to assess the scenarios in *The Limits to Growth* books for their validity, and most importantly, for their usefulness in informing us about possible actual developments over the ensuing decades. She did her research on *The Limits to Growth*, so as to understand the modeling technique and to be able to interpret its outputs and then compared the scenarios against real-world data.

After Gaya's research was published, she continued to follow her curiosity and delved deeper into what her conclusions mean for our common quest for higher global well-being. Because, as Gaya highlights in her work, the fact that there has not yet been a global collapse due to environmental reasons does not mean that the future is rosy. We are starting to see a stagnation in human well-being. My expectation is that this stagnation will turn into a decline unless there is truly extraordinary collective action to break from business as usual, as well as decision making as usual.

My intense wish is that Gaya's study is used to accelerate what needs to be done: bring down the human footprint, notably by reducing the climate footprint in the world of the rich, where the minority of the 1.5 billion people out of the 8 billion global population cause the majority of the greenhouse gas emissions. This reduction in our footprint, although unachievable without the rich nations sharply reducing their material consumption, does not need to be a reduction in our well-being. As Gaya beautifully describes in this book, it is in fact an opportunity for humanity to achieve greater well-being, in a way that can last. To let go of an unsatisfying and ultimately doomed pursuit of continuous growth at all costs and find true purpose in helping people and all other parts of nature thrive. We'll want to listen to her message.

Jørgen Randers

Professor Emeritus of climate strategy at the BI Norwegian Business School

and co-author of *The Limits to Growth*

Praise for *Five Insights for Avoiding Global Collapse*

This book by a young courageous thinker is a call to action to her peers to shift from the competitive, conflict-ridden, and unsustainable overconsumption culture, towards collaborative mutually enriching relational culture that aligns to what it means to be human. The future of our planet depends on the ability of each of us to embrace our interdependence on one another within the web of life within which we are inextricably linked. This book is a must-read for all those searching for how to promote a future we can be proud to bequeath to our Children's Children.

Mamphela Ramphele
Ph.D. Co-President of the Club of Rome

This clear-eyed, scientifically based yet extremely readable book is essential reading. Read it to learn in careful and systematic detail why the relentless pursuit of growth of monetary incomes, without considering its patterns and ignoring its inequalities, is leading humanity headlong into disaster. But read it also for hope: to learn that changing course is possible. There are feasible ways to fulfill human needs equitably, with respect for nature and the planet, which would save the planet, transform societies, and enrich all of us.

Jayati Ghosh
Ph.D. Professor of Economics at the University of Massachusetts Amherst, and member of the United Nations Secretary-General's High-Level Advisory Board of Effective Multilateralism

Researchers over the last 50 years have put tremendous effort into finding excuses for not taking planetary constraints seriously. Gaya Herrington discusses their common misconceptions, but also provides guidance on how to wrestle with the reality of living on a materially constrained planned. She shows that there is only upside in embracing this reality: it helps you navigate future rapids more effectively. Study Gaya's contribution because it may rescue you from the planetary blindness your professional education most likely fed you, or at least condoned.

Mathis Wackernagel
Ph.D. Founder and President of the Global Footprint Network

Gaya Herrington's research shows in hard numbers what many have felt intuitively: that we are experiencing a hinge point in human history, and the hour is late to take action. With a call for humanity to embrace responsibility, Gaya is asking for much. But she is leading by example.

Hunter Lovins
J.D. President of Natural Capitalism Solutions

This book is for anyone longing to understand how the personal connects with the big-picture systems that have brought us to this perilous point in our human story. In a voice that is engaging, honest, and often moving, Gaya Herrington weaves together science with insights and anecdotes that bring her potent conclusions to life. She describes what is at stake, but also what we stand to gain by going beyond growth and transforming antiquated systems: a world we want to live in.

Julia Kim
Ph.D. Program Director, Gross National Happiness Centre Bhutan

Contents

About the Author

Gaya Herrington received her first master's degree in Econometrics from the Liberal University of Amsterdam and her second master's in Sustainability from Harvard University. She is a member of the Transformational Economics Commission of the Club of Rome, a recurring guest lecturer at the Victoria University of Wellington, New Zealand, and works at Schneider Electric as Vice-President of ESG Research.

Acknowledgments

I am grateful to all the people that reached out to me when my research on *The Limits to Growth* went viral in the summer of 2021. Having just had a baby, I was unable to respond to each of them, but I read every word. It is no exaggeration when I say that, at times, a brief message of encouragement from a stranger was the only thing that kept me going with this book. It was an honor and pleasure to work with some of the Club of Rome members on their new report, 50 years after the publication of *The Limits to Growth*. The thought-provoking, insight-packed, at times heated, discussions have meant a lot to me. They made this a better book, and me a better person. I thank my husband Mike for taking three months off after our daughter's birth so that I could fully recover, and for arranging his 10-hour workdays to be home before 6 PM so that I could write in the evenings. Our society places an insensible burden on parents, but at least you made sure that we shared that burden fairly. Lastly, I am grateful to my daughter Ella for letting me read books and papers hanging over her head while breastfeeding, and for sleeping through the night since four weeks of age. I could not have written this book without you being as good a baby as humanly possible. I am looking forward to meeting the adult you'll grow into.

Gaya Herrington

1. Introduction

I am 41 years old. This means I grew up in a time of optimism. The popular interpretation of Fukuyama's book *The End of History and the Last Man* as a prediction of the end of events may have been a misrepresentation; still, I believe it was a sign of the times. For a moment in history, we thought we had it pretty much figured out.

I don't think that anymore.

In this book, I will describe how I came to the conclusion that we live in a moment of extraordinary historical relevance. What a unique now-or-never moment we have to turn around humanity's current trajectory towards something much better than the society we live in today. And why failure to make this turnaround will result in a significantly worse one.

I will advocate for deep systemic changes to our global society, although I should point out that I have benefitted more from our current systems than most people on this globe. I grew up in one of the richest countries in the world in terms of wealth per capita—The Netherlands. I studied Econometrics and paid off my student loans of less than EUR 3000 within five months after graduating cum laude in 2004. I ended up, as typically happened with Econometrics graduates in those days, in the financial sector. As it turns out, I had joined one of those companies that dealt exclusively in asset-backed securities, the financial products that are now known as the instigators of the 2008 global financial crisis (GFC). I left that company sooner than I had intended at the start, with a vague sense that something was wrong. A few months later, the GFC hit, but I do not claim that I saw it coming. At the time, I had not even discounted the possibility that something was wrong with me. Looking back, however, it was the first inkling that something had been missing in my economics classes. I made a career switch and led an activistic non-profit that advocated for mainstreaming organic and fair-trade products for a few years, at the time still convinced of the responsibility of private individuals to bring about change. Since then, the GFC had disrupted societies around the world, and I realized that it hadn't just been me who didn't understand what was going on in the financial markets. I went back to finance but on the regulatory side this time, at The Dutch Central Bank. After a few years there, I was approached by KPMG to work for them in New York City (NYC). They provided me with a high enough salary and sufficient downtime to actually enjoy the place. After three winters, however, the city had lost its appeal to me. I told my boss I needed to move to the West Coast, Los Angeles to be exact. This was no problem; KPMG has offices everywhere, and I was working remotely. The city of broken dreams made plenty come into my life. I finalized my thesis for a master's degree in Sustainability from Harvard, remotely in a small house overlooking the Hollywood sign. I met my now husband, who had also lived all over the world for study and work, including in Japan, Australia, and Germany, and several places in the United States (US). Unburdened by any debt, we bought a house in the Washington DC area. While pregnant not much later, I turned my thesis into a journal article, which went viral last summer. For several days, the headlines on major American news pages trumpeted that my research proved we are on the brink of societal collapse. A few days later, British pages touted the same headlines. Then, within a few weeks, I saw my name popping up in languages I do not know—from Swedish to Greek, to Chinese, to Sinhala.

So, I benefitted from publicly funded quality education in Europe. But when the time came to pay it back through taxes on the high salary this education allowed me to make, I moved to a place that doesn't tax incomes nearly as heavily. There, I was able to enjoy the unparalleled quality of the American education system, which stays out of reach for most American-born citizens. Ironically, the only reason I was even able to move to the US and build out my career was that major companies cannot find a sufficient number of educated workers domestically. I enjoyed the freedom of movement across the country and the world that globalization offers to skilled, white-collar workers. For me and my husband, the American Dream came true. But that doesn't mean the system works. I remember the image of a smoking homeless woman, clearly pregnant, in the subway in NYC. Or the view from my balcony in the Hollywood Hills of the smoke hanging in the San Fernando Valley after yet another one of those fires that become worse every year due to climate change, as firefighters have been warning for decades. Today, there are places in Southern California where water sometimes doesn't flow anymore from residential faucets, but you wouldn't be able to tell from the ever-green lawns in Beverly Hills. In my first year at KPMG, I called HR asking why they had stopped taking social security contributions out of my pay. The woman on the line told me I had hit the maximum contribution. "But that makes no sense", I said, "why would there be an income maximum on that?" She replied: "Ma'am, that question is way above my pay grade". I've thought about that answer more than she probably realizes. The part of my income that would bring me to what I think should be my social security contribution has been added to my charitable donations since then. But this changes nothing about a regressive system's structure. You could say that it is easy to advocate for more sharing and equality between people after I have benefited from this structure, when I have plenty to share. This wouldn't be entirely true; writing this book took a lot of time, energy, and dedication. I did it next to a full-time job while taking care of a newborn. I worked hard and sacrificed for my degree, publications, and career. Still, it is true that finding the time to write is easier when you're not working two jobs to make ends meet. Or to find the energy when you're not starving. It is easier to have headspace when you're not worried about your son being recruited by gangs. And that is precisely why I advocate for more sharing and equality between people. I cannot change this society by myself. But I am convinced that we have the knowledge, the capabilities, and the will to do so. And although this is now the biggest constraint, humanity still has time ... If we work together.

This book inevitably reads as written by a WEIRD person, since I am Western, educated, industrialized, rich, and democratic. I am conscious of the limitations this brings to my perspective, but I don't think it impedes this book's relevance. Although my commentary on our global systems is useful to anyone, it is most pertinent to readers in mature economies. The limits to growth are most visible there, and the main responsibility for systemic changes lies on the rich nations' shoulders.

An academic read of my research is still accessible on the site of Yale's *Journal of Industrial Ecology* (Herrington 2020). A simple online search will bring up some of the articles on popular news sites about my work. So, what is in this book that you cannot read in fewer words in those online articles?

First, I have updated my research based on data available in early 2022. This update was previously unpublished. Second, the understanding I have gained from these data has evolved through a combination of new global developments, my continued research into adjacent topics like inequality and well-being, and rolling insights from the numerous discussions I have had with people who have come my way since the journal's publication. I have given many lectures and speeches over the past year about my research and have since also been asked to join a Club of Rome (CoR) commission. The exchanges with audiences after my speaking engagements and the fellow commission members have sharpened my insights. Third, in contrast to the journal article that prompted media attention, I have added much more of my voice to this book. My 2020 journal article was mostly hard data analysis, but there is a non-quantifiable dimension around the *Limits to Growth* works that I have covered in this book. Academic standards dictate we stay neutral in our articles. This tradition is understandable and necessary; it is not the role of science to tell us which values to live by. But as I will lay out in this book, avoiding societal collapse is highly unlikely without a change in our societal values. It is justified, in fact imperative, that researchers like myself discuss values explicitly. The content of this book is supported by the latest research, of my own but also that of many others in a broad range of fields—from psychology to sociology, history, biochemistry, finance, and economics. Most of this research is from academia, but I am not an academic. Working at Schneider Electric, which held the title of the most sustainable corporation in the world in 2021 and is intent on staying a frontrunner in that field, I have more freedom to discuss values explicitly. So, I thought I would use this to offer up values and a new goal for humanity I think are worth fighting for. And to call out which current stories and societal goals are not worth killing or dying for.

This book has been peer-reviewed and contains enough details on my research for those that would want to replicate it. Nevertheless, readers that prefer to skip the more detailed and technical parts, such as the data source descriptions, should feel assured that they can do so without losing the book's main thread. Similarly, sustainability experts, systems modelers, or economists may find there is little new for them in those paragraphs pertaining to their topic of focus. These different fields are joined together in my insights, however, with the *Limits to Growth* works acting as the connective tissue. And these insights are relevant to anyone. Each covered separately in a chapter, they are summarized as follows:

Insight 1: We are connected, and acting like we are not has led us to the brink of collapse. An interconnected world is messy but also rich. Trying to control our global system as if it was a simple collection of isolated parts will turn things from messy to ugly. This is what we have done with our global ecosystem and society.

Insight 2: Growth is not a good goal; in fact, it is the cause of society's problems. The pursuit of expansionary growth was identified decades ago by scientists from the Massachusetts Institute of Technology (MIT) as the root cause of the world's persistent problems such as poverty, conflict, and pollution. They prophesied that persisting in this growth pursuit would bring us to where we are today.

Insight 3: We need to fundamentally change society's priorities if we want to avoid significant declines in our current levels of well-being. My research indicates that global growth will halt, one way or another, in the medium term. Avoiding a subsequent steep decline in overall economic output and welfare levels requires the global society to adopt a new goal of well-being and respect for Earth's limits, with frameworks that put services such as education and health care at their core.

Insight 4: Time is of the essence to make this change. The MIT scientists' warning has, by and large, not been heeded because it went against the prevailing economic and political thought. As a result, we are now fast running out of time to make the necessary systemic changes. Humanity needs a sustainability revolution. Such a transformation can only be achieved by adopting a new shared narrative of who we are, what world we want to live in, and what unique role we get to fulfil in that world.

Insight 5: The end of the growth pursuit does not mean the end of progress; quite the opposite. A relentless pursuit of growth is not only ecologically degenerative but also socially. Social organization around a concept of dynamic equilibrium rather than relentless growth comes naturally to humans. A society shaped around a narrative that ascribes purpose to the fostering of human and ecological well-being can still allow for growth but only to the extent that it leaves us happier, as our definitions of success and prosperity will have matured far beyond the mere avoidance of collapse.

What exactly would such definitions and our general ways of doing things look like in this global equilibrium? In Chapter 7 of this book, I will attempt to touch on a few of those aspects. This last full chapter (Chapter 8 only contains a comparatively short conclusion) also includes a core summary of a new book by the CoR. Published exactly half a century after the publication of *The Limits to Growth,* based on which I developed my research, this new CoR book pinpoints the levers in the global system that we should pull to trigger the sustainability transformation we need.

But of course, the above insights raise many more questions than I could ever go over, from the global to the local, from concept to practice. If society's goal was not growth but enough for each, what geopolitical shifts could we expect? How would unpaid care work be included in our economic measures? Would the cultural norm of sleep deprivation shift to a midday nap or shorter workdays? In what currency would our income be paid? And would this income still be higher for marketing executives than for teachers or nurses? Would multilevel marketing still be legal? How would our justice system be transformed? How would the entertainment industry change? Or the sex industry? Would we thank our mothers for birthing us as customarily as we thank veterans for their service? Would the military and police include a "forest infantry" or "animal squad" to protect other life forms to the extent that some material objects are protected now? How would our attitudes towards psychoactive fungi and plants change? Would the majority of mammal biomass on the planet still be from animals we hold for consumption? There is no limit to the number of questions about how society would be transformed once humanity embraced its purpose to safeguard life in all its forms to thrive on this planet indefinitely.

This is just as well; the most useful information helps you ask the right questions rather than providing answers. My hope is that this book will help frame the right questions for you and gives you the courage to seek the information you need so that you may become part of a movement toward a world in which humanity has become far more ambitious than just seeking ever-more, and instead, is striving for better.

2. Systems Thinking: Everything Is Connected

2.1. Systems Thinking

The lungs of the Earth have reversed behavior. For the first time in its existence, the Amazon now emits more CO_2 than it absorbs due to "large-scale human disturbances" (Gatti et al. 2021). We are living in the sixth mass extinction of life on Earth, and partially because of our encroachment into wildlife habitats, 2020 brought us a seemingly sudden pandemic that completely disrupted our already feeble sense of normalcy (Tollefson 2020). What many of us initially thought would be over in a few months, lasted years. Even COVID-19 could not stop an overdue reckoning of our racism from finally reaching its watershed moment, however. And at the time of writing, Russia's war in Ukraine has us heading into winter with renewed anxieties around geopolitical and energy security. Our world is full of tipping points, counterintuitive conjunctions, and inertia. That makes it considerably challenging to manage already, even if we were unified around the same goal of a thriving planet with enough for each. But if the disappointing COPs, erosion of democracy, and vaccine conspiracy theories of the last few years gave us anything, it was more evidence that we seem unable to come together even to save ourselves.

How did we get here? In this book, I will explore what I believe to be the root cause of our current social, environmental, and economic problems. But before I can do that, we need to align on what "root cause" means. And for that, we need to talk about systems. This first chapter will introduce some of the basics of systems thinking. It is not meant to serve as a college course-grade introduction to the subject but should give you enough background to appreciate the way of thinking and modeling that underlie the *Limits to Growth* (*LtG*) books, which I will introduce in the latter half of this chapter. If you would like to learn more about systems thinking, I encourage you to read some of the excellent online papers and books, a few examples of which are the free online introduction from Daniel Kim (2021), the management book from Peter Senge (1994), the socially focused book of Irl Carter (2017), or the foundational and analytical treatise by James Miller (1978). Of course, my personal favorite is the sustainability-focused book *Thinking in Systems* by Donella Meadows (2012), one of the *LtG* authors. The *LtG* books were based on a system dynamics model, and the authors were obviously pioneers in systems thinking.

When it comes to understanding the world around us, systems lie between deterministic and random processes. Randomness is a part of life, although you probably don't often consciously encounter pure randomness during the day (unless you work in a casino or your job involves a lot of radio communication, in which case you'll be picking up atmospheric noise). On the other side of that spectrum are the deterministic processes that we can describe with formulas. Sometimes, it can take a lot of intelligence, time, and effort to find these formulas, but once you do, they will keep working because the relationships you studied are fixed. It doesn't matter what story you tell about the pendulum, or if you surround it with a lot of other pendulums. The pendulum always swings the same way given the gravity force, mass, arm length, angle drop, friction, and air resistance. But there is an area in-between those two extremes of random and deterministic events, and that is systems behavior.

You are a system, and your behavior is an example of a system's behavior. We are built from other systems, like our digestive system, for example, or a single organ. We live in systems: our family, the organization we work for, our country, global society, and our ecosystem. The essence of systems thinking, in one word, is interconnectedness. In a few more words, it is the recognition that, to understand a system's behavior, you cannot just study its individual parts in isolation; the total of relationships between the system's parts—its structure—is often just as important. Why? Because once parts start to interact, they create a meta-behavior, the behavior of the system itself. Thus, a system's behavior is not just the sum of the behaviors of individual parts and, therefore, cannot be fully predicted. But it is not random either. And it can get tricky. Because a system's parts interact, they change the behavior of other parts of the system to which they are connected. For example, a deck of cards is a collection, not a system. Their values do not change, and if you add another card to a card collection, the new total value of the collection is simply increased by the value of the added card. If we now play a poker game with those cards, they have become part of a system. That is because we added players (people), rules, parameters (the card values), and a goal (e.g., to end up with the most total value of cards without folding). Now, the deck of cards has become much more interesting because, by making it part of a system, the impact of adding a card anywhere has become much harder to predict. This is not just because it has become more complicated by adding rules. Complication means it takes more time to analyze something, but once you have completed this analysis, you have your answer, and it will stay correct. In other words, it may take more time and effort to predict the future in a complicated situation, but once you come this far, it is smooth sailing from there on. An example is the mathematical formula for the trajectory of the aforementioned pendulum. Adding a cord with a weight to the pendulum will complicate the formula, but with enough time and effort (given the right knowledge), you can find it. In the case of the poker example, in contrast, the situation has become more complex: The right move constantly changes based on how the parts in the system are changing. They depend on moves from other players, which, in turn, depend on the moves and the expectation of the moves of others. In such a complex situation, sticking to a previously right answer is often a losing strategy. Adding humans to a situation typically has this complexifying impact, which is why, as a rule, it is not a good idea to assume humans can simply be added to the situation like they are extensions to a collection of parts. We see this in broader society too. We live in the times of Big Data; we have more information available than ever before. But has the world become more predictable? Of course not, quite the opposite. It makes the world less predictable, because we are changing our behavior based on the available information. Any insight derived from a new piece of information that triggers a change in behavior (and is a finding really an insight if it does not trigger any change?) causes others to respond to that behavioral change, thus very quickly rendering the information of the previous insight outdated. This is also one of the reasons why the financial markets have not become more stable with all the price information available to us, despite "efficient market" arguments. But I will come back to finance and economics in later chapters. What it comes down to is that more information feedback means more connections between a system's elements. The more interconnectedness in a system, the more complex and thus dynamic it becomes. This does not necessarily mean things become less stable, or more stable for that matter; dynamism and stability are neither necessarily mutually exclusive nor conducive. Complexity can make a system more resilient to shocks, but it can also make it collapse under its own weight. Trying to predict and control amidst high complexity, however, is a great way to get yourself into a lot of trouble fast.

Interpreting outcomes in a system requires a different—and arguably a more nuanced—mode of thinking. For example, if our society was complex rather than complicated, the existence of systemic social inequalities such as racism and sexism, as well as homo-, trans-, and enbyphobia, would not be disproven by simply pointing at a successful female, black, gay, trans, or gender non-conforming person. Systems are not fully deterministic, so there will be outcomes that are contrary to the way the system is set up. Nevertheless, the existence of Oprah Winfrey and RuPaul does not prove that climbing the societal ladder is a process with merit as its sole determinant. Because systems are also not fully random either, we can still learn to understand them to some extent. Although systems are not fully predictable, we can become better at dealing with them, and the first step is to adopt systems thinking. Without understanding systems, we tend to focus on the parts of a system's behavior that are the most visible: the endless stream of events. But there is a whole systems iceberg below that waterline of visibility.

2.1.1. The Systems Iceberg

We all observe events, the part of the systems iceberg above the water level (Figure 1). Events are easily spotted, which is why they receive a lot of our attention and are relatively easy to quantify. However, events produced by a system form some pattern, which can be partially explained by the system's structures. Delving one level deeper, below the structures, we find mental models. Not the econometric models that I was taught to build but the qualitative views of our world. We could call the collection of our mental models a paradigm: the "often-unconscious notions we hold around how things work" (Kuhn 1962). A paradigm shift, then, changes how we think the world works, and as a result, it also changes how we behave. But changing someone's mind on this level is, as we all know, difficult. It is possible, but it requires a compelling vision, meaning a story of how and why things should be, and how to get there. That is the generative level, which is to a large extent impossible to quantify but with a tacit power to change the world.

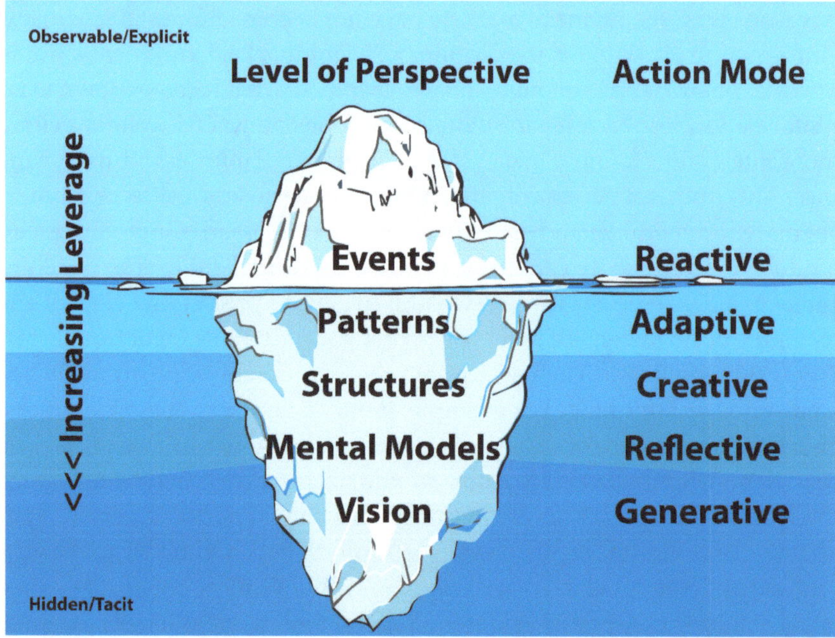

Figure 1. The Systems Iceberg. Source: Figure by author, graphic design by Hilary Moore.

Each level in the systems iceberg holds a certain potential for impact. The deeper you go, the larger that potential becomes. We can really only react to individual events. A school shooting, to give a dramatic but well-known and intuitive example, leaves few options other than to hide, run, or, in a rare case, attack the shooter. One level deeper, we can detect a pattern and perhaps adapt, for example, with the active shooter training courses provided in the United States (US), where the vast majority of mass shootings happen. This is not reactive because no shooting is happening at the moment the course is taught. The course is adaptive in the sense that it is aimed to prepare for a future event now that we have detected a pattern of regular mass shootings in the country. Then, there are the structures, such as legislation, other rules, natural laws, and social relationships. Structures can be of various natures, biological or artificial, tangible or virtual. Gun legislation is part of the structure in this example, and so is the school and the class the shooter was a member of. This level already seems quite a bit more powerful than the reactive one, and it is. But acting on a system's structures is still only creative. Reflection on an event, pattern, and structure is a much more powerful way to affect change than the creation of, say, new laws or quantitative models. To stay with our example, although gun laws might make a difference, there are countries with relaxed gun laws that do not have school shootings, certainly not with the frequency that they occur in the US (Kellner 2008). Scholars have been pointing out for a while now that the US needs to address this issue at the level of the mental models, specifically, notions surrounding what constitutes manhood and what role violence plays in that identity (e.g., Katz 2012; Kellner 2008). The fact that this discussion has been largely shunned so far ultimately comes down to our shared narrative about what values such as freedom, safety, and self-reliance mean. Our societal priorities flow from that vision because it provides us with an image of a world where such values have come to life. A societal system built on a vision in which freedom includes being able to defend oneself with a firearm, will look different from the one in which freedom is defined more along the lines of living free from lethal danger. In general, at this vision level, a new compelling narrative can generate major societal change. Which is probably why history seems to indicate a disproportionate risk to life of being a visionary.

2.1.2. What Does It Mean That We Live in Systems?

There are two major implications of living in systems, both philosophical and practical at the same time. Firstly, it means you are not completely in control of your own destiny. This is an unsettling thought for some. Your innate abilities and personal values will make a difference in your life. I believe that empathy, courage, and discipline will pay off in the long run for almost anyone. However, these behavioral qualities are not the sole determinants of life outcomes. Systemic forces play a major part in what different people can accomplish with the same efforts. This is anathema to the popular notion of individual responsibility, so people who are emotionally invested in that mindset may feel uneasy when being introduced to systems thinking. Admittedly, thinking in systems can be quite overwhelming sometimes. If everything is connected, where does one start to make any change at all? And how, if it is not as simple as pressing a button? There is a way to still make a difference, but not with force. Influence, rather than strength, is the key to making a lasting impact when working in a system. According to renowned international lawyer and foreign policy analyst Anne-Marie Slaughter (2017), young people and women tend to understand interconnectedness faster than older men. I suppose the reason is that younger generations and women are simply more used to this form of power, rather than having much coercive control. On the other hand, if you're used to achieving things through hierarchical command-and-control structures, having to think and work in a system can be deeply frustrating. Based on my experience, it can even lead to a small existential crisis. One time, when I was presenting on interconnectedness and systems thinking at a large financial multinational, an attendee stood up in the middle of the meeting (and my sentence) and walked out. My colleague went after him and told me later that the person had spent the rest of the scheduled time in the smoking room fuming about how my presentation was a waste of time. "My body is a system . . . .", he apparently had said while taking another inhale from his cigarette, "What the hell does that even mean".

Secondly, living in systems means you are not alone. That is the other side of everything being connected. This is not a new notion; the reason this statement reappears in so many new-age books is that it is a common principle in many traditional beliefs. Interconnections in a system can either work in a reinforcing or counteracting (balancing) way. As Donella Meadows (2012) points out, by far, the most feedback loops are of the latter kind. In the background, interconnectedness is constantly working to keep our world balanced. If you care about something, you are never truly alone in your efforts to protect, nurture, or defend it. I mean this in a practical sense. Once you move beyond the feeling of disempowerment and open yourself up to behold the system on which you are focused at that moment, you might gain an understanding of it. Once you do, you may be able to use its interconnectedness, perhaps with the tools in the next section or the last chapter of this book, not only to get things done but also to achieve things you didn't even imagine you could be part of.

2.1.3. Systems Thinking Tools

We will become a bit more practical on systems for now, after which we are all set to dive into *The Limits to Growth* and my research. Systems consist of three things: elements, interconnections, and a purpose. A system's purpose is almost never directly visible. But it is the most powerful part of a system because it resides at the vision level of the iceberg. If you want to know a system's goal, observe its behavior. Ask a company or government committee what the purpose of their cooperation is, and they will give you their mission statement. Observe their behavior, however, and sometimes you might find quite a different goal. Elements, on the other hand, are the easiest to see. They consist of stocks and flows. Your cells are the elements of the system that is your body. An example of an economic system's stock is accumulated wealth or assets, while a flow is an income such as your salary or a country's gross domestic product (GDP). Flows are determined by rates. The birth rate determines the number of births for a given population stock, for example. The interconnections are the relationships between the elements. These are less easy to see than elements but still observable, and they are often called feedback loops. Feedback loops can be positive or negative or, alternatively, reinforcing or balancing. Some of these interconnections are also flows, but oftentimes, feedback is an information loop. The sensors that are part of smart buildings and the Internet of Things, in general, form a modern example of feedback loops. Sensors in a water delivery system, for example, may detect a reduced water flow in a certain subsection significant enough to trigger a warning for leakage. This information is relayed to a repair crew that repairs the leak, restoring the water flow to within normal parameters. The information feedback loop through the sensor helps maintain the stock of available water while keeping its cost and waste (two other flows, financial and physical) as low as possible. Elements and relationships can exist at the event, pattern, structure or even mental model level, although I have most often seen them used to depict a system's structure.

2.1.3.1. Stock & Flow Diagram and System Dynamics Modeling

One of the tools in systems thinking to visualize the elements and interactions of a system is a stock & flow diagram (SFD). An SFD is exactly what it sounds like: a depiction of how the stocks and flows interact, including the information exchanges. System dynamics (SD) modeling could be called a formalized version of an SFD. An SD modeler will assign numbers to the system's interacting stocks, flows, and rates. Thus, the system's interactions are quantified but not to make point predictions. Systems thinkers do not presume they can predict the future this precisely in the first place. Rather, by varying these rates, SD models can produce different scenario runs, which can be useful to understand the general dynamics of the system. This kind of scenario analysis can enable us to prepare for a future in a complex environment. MIT professor Jay Forrester (1971, 1975) is generally considered the founder of SD modeling. He is also the one that created the first SD model of the entire world, World3. This was the model used in the research supporting the *LtG* books and will be discussed in more detail in the next section. We do not build any new systems in this book, just examine existing ones. For that purpose, a tool that looks at a system from a higher level might be more useful. That tool is the causal loop diagram.

2.1.3.2. Causal Loop Diagram

Causal loop diagrams (CLDs) provide an overview of the relevant interactions in a system. A CLD does not necessarily differentiate among all the parts of the system but rather focuses on the system's general behavior. Causal links are still either reinforcing or balancing and are indicated with a plus or minus sign. The balancing causal links can also be depicted as dashed lines. Population is often used as an introductory example of a CLD, such as in Figure 2 below. The stock is the population number, let's say, people. More people mean more births at a given birth rate, so that is a positive causal link or one moving in the same direction. More births mean more people, so that is another "same" causal link. Together, they form a reinforcing feedback loop (shortened to "R" in the figure). More people also mean more deaths, but more deaths mean fewer people. So here, we have one positive causal link and one negative link, which is also said to be moving in the opposite direction. An opposite and same link together form a balancing feedback loop (shortened to "B").

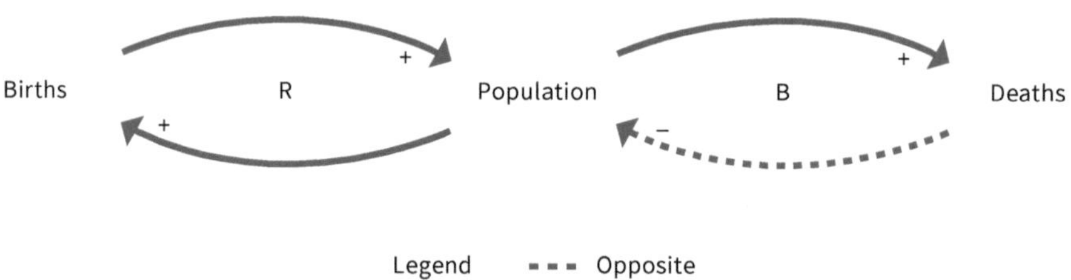

Figure 2. A simple CLD for population. Source: Figure by author.

2.1.3.3. Delays and Tipping Points

The above building blocks are simple enough; put them together, however, and soon, the system starts to display all kinds of dynamics. Delays, for example, are common in a system. A delay is when an effect from a cause takes time to materialize. Stocks often work as buffers, which can cause delays in the materialization of effects. In a CLD, a relevant delay is typically depicted with two dashed lines in a link. The bathtub is a common, simple example. If the bathtub is large and mostly filled with cold water, it will take time before adding a flow of hot water makes the water temperature acceptable. It might not even be possible without reducing the cold-water stock. If you have a barely funded savings account, even a small withdrawal can put you in the red immediately. Several small withdrawals from a large stock of savings, however, will not impact your wealth much, unless they continue over a long time. That is why in large systems consisting of many stocks, change takes time. As Donella Meadows (2012) mentioned in her book, this is important to know if you work in system change because this realization will help you to persevere. However, it can also lull you into a false sense of security. Our ecosystem is very complex and consists of some unimaginably large stocks. Even so, they are still finite. Humanity filling the atmosphere with ever more CO_2 will not show much effect at first because of the ecosystem stocks that work as buffers. Oceans, for example, have absorbed around 30–40% of the CO_2 and 93% of the heat added to the atmosphere through human activity since the start of the industrial revolution (Heinze et al. 2021). Without the ocean functioning as a buffer, the scale of atmospheric warming would already be much larger. Permafrost only starts to thaw after having absorbed increasing heat over an extended period. But these, and many other buffering effects, have their limits, which is why climate scientists have been warning about "climate tipping points" (e.g., Lenton et al. 2019; Intergovernmental Panel on Climate Change 2021).

Tipping points are a common dynamic in systems. A tipping point is that moment when a series of small changes has built up to a level where a sudden large movement is set in motion, and something in the system permanently changes. The system does not revert to its initial state, even if the drivers of the change are fully removed. The buffers that were present in the system might change their behavior. They might partially or fully stop buffering, like with a full bathtub or the ocean at the climate tipping point. Buffers can also change function and become a reinforcing loop, such as a bank account turning from a buffer to adding an outcome flow because of the interest charged on your debt, or thawing permafrost releasing its previously stored carbon into the atmosphere, turning it from a carbon sink to a carbon source (Natali et al. 2021). Social examples include the #MeToo or Black Lives Matter (BLM) protests. Once the proverbial dam breaks or bucket overflows, the unfolding event(s) like viral tweets or worldwide protests are easy to spot. We can quantify them to a large extent: tweet counts, likes, and estimates of the number of people who attended this many rallies on these dates. But they had been building for a long time under the surface of the iceberg. And society has not been the same since.

2.1.3.4. Leverage Points

One great way to use interconnectedness is by working in leverage points. These are points in the system where your efforts will be amplified by virtue of its structure. Small changes in a leverage point can generate large changes in the system's behavior.

First, there is the level at which you work. As already mentioned, the deeper you go down the iceberg, the more impact you can make. Let's look again at Figure 2, for example. Births and deaths are the events. Visible and important events in an individual's life, but each a small part of the general population patterns of reproduction and mortality. The CLD works at the pattern level. It shows the general dynamics of the population, and we could quantify it with population surveys. If we want to predict the future population, we build a model. Many organizations around the world build and use such population models. Such models contain assumptions on death rates and the average number of children per woman, among other things, which are based on historical averages and expert opinions on these rates going forward. This is where our mindset comes in, which, in this book, I will define as referring to both the prevalent mental models and the narrative. What social and cultural factors could influence a birth rate? What mindset influences the development of policies that prioritize health enough that life expectancy increases, and the death rate declines? Cultural notions surrounding the definition of a woman's role, economic factors co-determining her career options, institutional support for new mothers such as access to health care and birth planning, and subsidized childcare all play a role. Such notions have significantly differed over time and space, and birth and death rates have changed with them because these notions are the most influential in shaping population dynamics.

Second, on each of the iceberg's levels there will still be points with more effect than others. Once we have formed an overview of the system, most often at the structural level, we can start to identify at which points in the system we can expect to have more influence than others. In Chapter 7, for example, I will discuss sustainability leverage points for the world in 2022. Identified using a systems tool, these points hold the potential to trigger the transformative changes in the global system that society needs.

Sometimes, I find it makes sense to speak of root causes instead of leverage points. Root causes are leverage points, but additionally, a root cause signals a causal chain of events that does not always apply to a system. If everything is connected, sometimes it just doesn't make sense to speak of one thing coming before the other (think of the classic chicken or the egg question). But in some situations, it is insightful to follow the direction of a system's flow for a while, such as in the case of a corporate and government policy setting. Not uncommonly in policymaking, the focus is put on the unwanted symptomatic event instead of addressing a problem's root cause. The unwanted event is almost always more visible and draws attention because, well, it is unwanted. So, the policy attacks the symptom. Some people thought banning homelessness would make it go away, for example. There can be several reasons for this misplaced focus. There may be pressure for fast results. Perhaps the manager or politician does not have much systems understanding. Or maybe they do, but they know their constituents do not. Or all three reasons together. Either way, this misdirected focus is a major reason why so many government and corporate policies fail (Mueller 2020). To some, saying "you complained about this thing and see, I attacked it!" may look like decisive action taking. But if the policy did not address the root cause, the issue will most likely return in the medium to long run. At best it is a waste of resources, and sometimes it even makes the situation worse. By then, however, the bad leader might already have retired. Tragically, root causes are often not that visible. At best, results become visible after some delay. But not seldom, successfully attacking root causes means preventing an undesirable symptom from ever happening in the first place. There has been a major impact on the system's behavior. It might even have been prevented from breakdown. But it is not easily observable. Root-cause policymaking is good leadership but only makes an impression on the very attentive shareholders or voters.

2.1.3.5. Archetypes

Becoming smart with systems will enable you to discern recurring behavioral patterns. These are called archetypes. Recognizing archetypes allows one to anticipate future dynamics and so avoid common mistakes and risks. There are many more archetypes than I will discuss here, but one example of an archetype is the intuitively clear *Fix that Fails*, depicted in Figure 3. It is the behavioral pattern that can occur when a problem has both a fix and a solution. The fix is easier (e.g., cheaper or faster) but only effective in the short term, and it comes with unintended side effects that at first may not be obvious. These side effects make the problem worse and/or diminish the system's receptivity for the long-term solution, possibly resulting in more fixes.

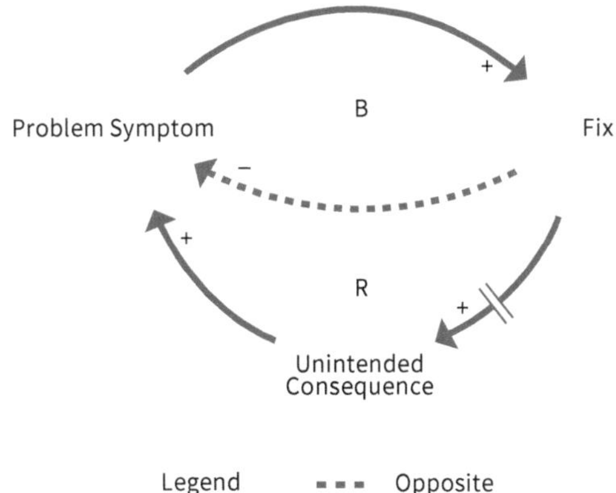

Figure 3. Fix that Fails Archetype. Source: Figure by author.

Everyday examples of this archetype abound, as do more tragic ones. You are tired because you didn't sleep enough, so you fix that with coffee. It boosts your energy for a while, but the effect does not last long. The actual solution, of course, is to sleep enough. Especially if you use this fix in the evening, however, drinking coffee might keep you up for longer and disrupt your sleeping pattern, leaving you more tired. A more sensitive example is spanking. Hitting children as a means to change their behavior is unfortunately still common. But research very convincingly shows that spanking does not work (Smith 2012). Yes, in the immediate term, the child will stop their unwanted behavior when faced with violence, but the spanking fix will not deter the child from the same behavior in the future—quite the opposite, in fact, because the unintended consequences of hitting a child include increased aggression, antisocial behavior, and sometimes, lasting mental health problems. That is why the corporal punishment of children has been recognized as a human rights violation since 2006. Another example of a *Fix that Fails* is people's quest for happiness. Hormones are one of the factors that influence our life satisfaction (e.g., Dfarhud et al. 2014). Three of the many hormones that our body makes are dopamine, serotonin, and oxytocin. All three make us feel good, but they differ in how long their effects last. Long-term happiness is supported by serotonin- or oxytocin-inducing activities like cuddling, other types of skin-to-skin contact, and meditation or prayer (Lustig 2018). The effects of dopamine, released during activities like shopping or reading a social media post, are short-lived. Moreover, research shows that pursuing dopamine boosts leads the brain to downregulate; it becomes less sensitive, not just to dopamine but also to serotonin and oxytocin. Tragically, however, studies also indicate that many of us are subconsciously seeking that dopamine fix, with the unintended consequence of becoming less sensitive to the hormones that keep us genuinely happy.

Another example is the *Success to the Successful* archetype. This archetype occurs when people are competing for a limited resource, and the chance of success in obtaining more of the resource increases with one's current level of success (Meadows 2012). The general archetype is depicted in Figure 4.

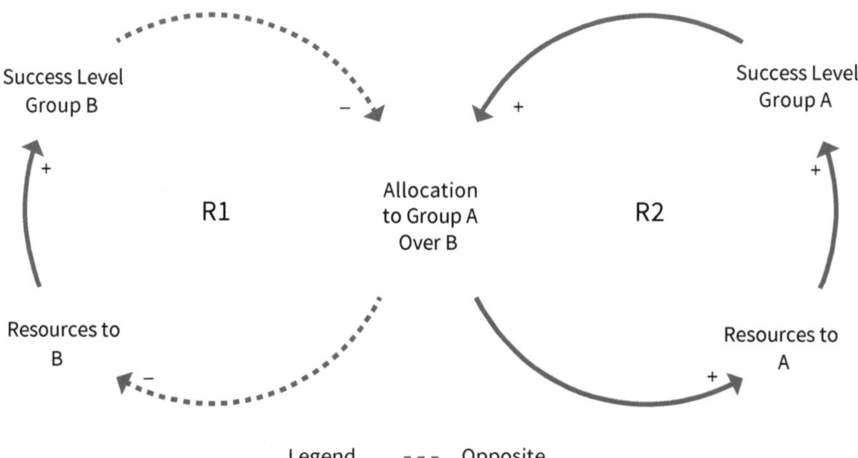

Figure 4. The Success to the Successful archetype. The numbered Rs indicate that there are two reinforcing feedback loops. Source: Figure by author.

One example of this archetype is the positive correlation between high earnings and health status and education level. This may make intuitive sense in a society where education, nutritious food, and access to medical care when necessary are expensive. Good health and a high education level, in turn, are also major predictors of higher earnings. This example is depicted in Figure 5.

We will come back to this archetype in Chapter 5 because it plays an integral part in the economic, social, and environmental inequalities observed in the world. In that context, a key insight from this archetype is that there really is only a very small initial difference necessary at the start to grow into staggering inequalities over time. The *Success to the Successful* effect is compounding, meaning that, over time, any difference will grow exponentially. The ground-breaking 1990s computer model Sugarscape by researchers Joshua Epstein and Robert Axtell (Epstein and Axtell 1996) is an illustration of this effect. Sugarscape was a virtual society set in two mountains of sugar with individuals, "agents", that needed sugar to survive. There was some variation in needs and abilities; some agents needed a bit more sugar than others, and some were able to move a bit faster or see farther. Depending on the conditions inserted into the model, all kinds of familiar forms or self-organizing dynamics appeared over time, including the formation of societal hierarchies, periods of warfare, peace time, migration, and trade. Another key phenomenon that arose was wealth inequality. At the start of a Sugarscape run, agents were randomly allotted sugar. There was some variation, but like their innate abilities, the differences in initial sugar endowment between the agents were low. Over time, however, these small differences could grow into enormous inequalities, with some agents ending up with vast sugar wealth, while others were barely getting by on a subsistence level.

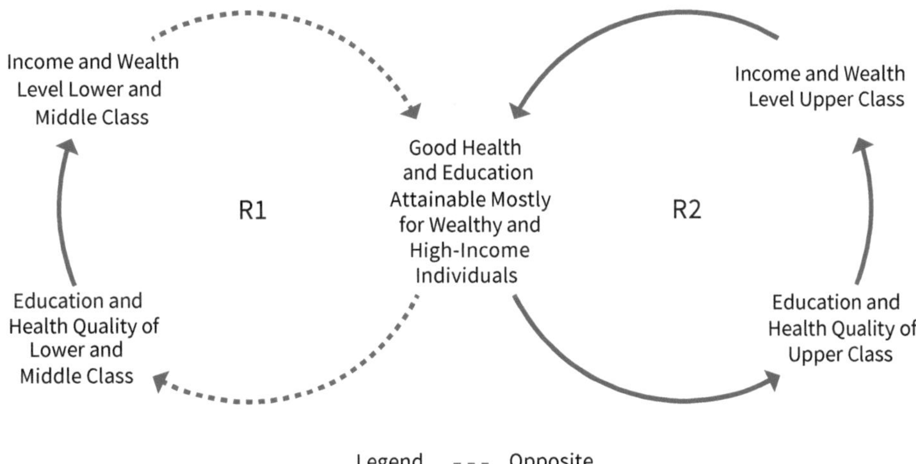

Figure 5. The Success to the Successful archetype for education and health. The numbered Rs indicate that there are two reinforcing feedback loops. Source: Figure by author.

People in general do not have a great intuition for compounding effects. When faced with the stupefying inequalities in the world, some can only conclude that this must be the natural order because, surely, a difference this large must exist for good reason. For example, popular author Yuval Noah Harari (2015) mentions in his book *Sapiens* that we simply do not know what causes today's gender inequalities. Without in any way suggesting that these differences are natural or just, he argues that the difference in bodily strength between men and women could not possibly be the cause because this difference is much smaller than today's inequalities between the sexes. It is true that we do not know, and the world certainly is much too complex for it to be explained by just one systems archetype. But it is not necessarily implausible for a small difference, in a time when physical strength was still more of a determinant for power, to grow into enormous inequalities over a longer period, even while what constitutes power evolved over that period too. A more recent simulation on gender inequality illustrates this. In the virtual workplace NormCorp, female employees were given only slightly less credit for successful projects than male employees, and slightly more criticism for failed ones. Such often unconscious biases have been shown to be ubiquitous in real-world settings, including workplaces (Nordell 2021; Turban et al. 2017). The cumulative impact of the credit/criticism bias over time resulted in a major gender imbalance in NormCorp. After 20 runs, a meager 3% difference led to a leadership tier that was 87% men. If sexist perceptions are remnants from a time when men really did have a biological edge over women in terms of labor productivity, thanks to more muscle mass, these small differences would easily serve as self-perpetuating outcomes, without having any basis in modern-day life.

Now, these models are interesting for studying inequality dynamics, but they don't offer guidance on the level of right or wrong and certainly no desired way forward. Models cannot do that by themselves because they are tools. For example, one could argue that simulations like the one discussed above support the notion of the natural order of inequalities. After all, don't they indicate that inequalities will always sooner or later arise as a regrettable but unavoidable phenomenon? That this is just how societies end up, even the simple virtual ones like Sugarscape, by hoarding and fighting each other over resources? But this behavior is the result of the system's design, and not just at the structural level. Aside from some innate abilities, Sugarscape agents were also given a purpose. They were each governed by a simple rule: "look around as far as you can; find the spot with the most sugar; go there and eat the sugar". One wonders how their behavior would change if their purpose was replaced with a different one, say: "look around as far as you can; find the agent with the least sugar, give them some of yours".

2.1.4. Limits to Growth Archetype

Another common archetype is *Limits to Growth,* which is shown in Figure 6. This concept is quite intuitive to most adults. We are not surprised when initial high growth runs into a source of resistance sooner or later, leading it to slow down. A forest fire spreads quickly, then dies out when the wood starts to run low. An innovative new product is adopted "like wildfire", but its sales level off as the market becomes saturated. Put in a more generalized systems thinking terminology, the *Limits to Growth* principle is the dynamics of the changing and often delayed diminishing forces that counteract an expansive force (Senge 1994).

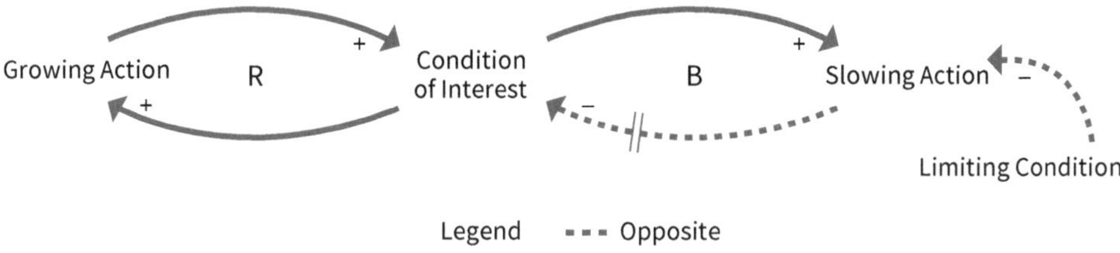

Figure 6. Limits to Growth archetype. Source: Figure by author.

This archetype is often observed in a population stock, which will initially increase rapidly until it starts to approach the carrying capacity of its environment. *Limits to Growth* is at work, for example, in what ecologists describe as "overshoot and collapse". This term describes a three-phase pattern observed in populations, which is depicted in Figure 7. In the first phase, a population is growing at an accelerating rate, because of the positive feedback loop described in the earlier population example; more adult members in the population mean more babies, and more babies mean more adult members in the future. At some point, however, the population starts to approach the number that can be sustained by its environment. The population may grow beyond this carrying capacity, but it can only remain there temporarily. Beyond the carrying capacity, population members do not thrive. This marks the start of the second phase, at which the population is said to be in overshoot. Because an increasing share of the population is too unhealthy to successfully procreate or survive at all, population growth slows down and subsequently ends altogether as the mortality rate rises. In the third phase, the mortality rate surpasses the birth rate, and the population starts to decline rapidly. Because of the steepness that typically marks this decline, the third phase is called a collapse.

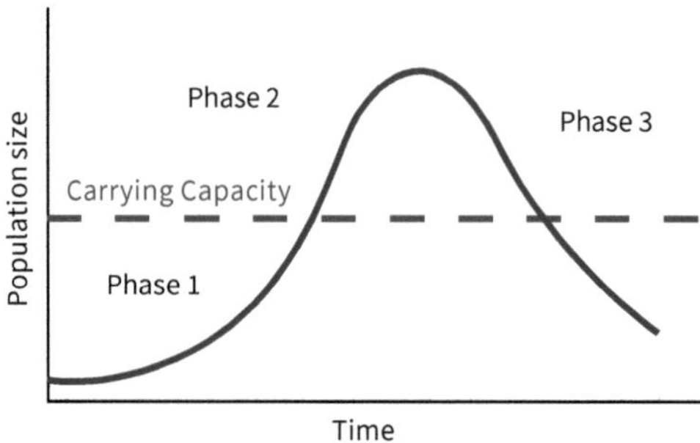

Figure 7. Stylized version of an overshoot and collapse pattern. Source: Figure by author, graphic design by Hilary Moore.

A growth process does not have to end in collapse. Logistic growth, depicted in Figure 8, displays a smooth approach to carrying capacity. Also called the S-curve, this pattern, too, is often observed in population dynamics and natural phenomena in general.

The principle that an expansive growth will be brought to a halt by some limiting factor, followed by a collapse or not, can be generalized beyond product sales, forest fires, and population. In fact, it applies to any real-world growth process (Chichakly 2009). This comes naturally for most of us because, once you start to think about it, you realize the *Limits to Growth* principle is balancing us and stabilizing the world we live in every day.

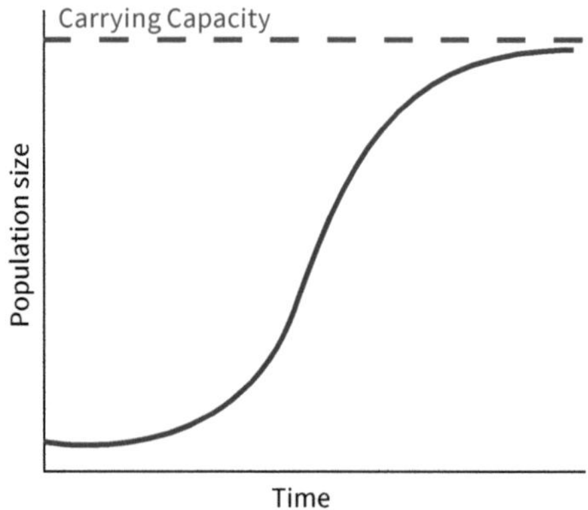

Figure 8. Stylized version of logistic growth. Source: Figure by author, graphic design by Hilary Moore.

Could economic growth be an exception? Given the dominance of the growth imperative in modern-day debates, you might think so. But the *Limits to Growth* archetype can be applied to economic activity as well. For example, economic activity creates pollution and uses natural resources like potable water, arable land, and fossil reserves. Let's take the environment's ability to absorb pollution resulting from economic activities. Nature can function as a pollution sink, but it can only absorb so much. With a young economy that is low in industrial activity, absorption limits may not yet be met, so economic growth increases standards of living (or "well-being", or "welfare levels"; these terms are used interchangeably throughout this book). With continuously rising well-being, pollution keeps rising too. At some point, pollution levels will start to come close to the environment's maximum absorption capacity. This is when the balancing loop kicks in, for example, through impacts on human health, the ability to grow food from polluted land and water, and the diminished availability of still uncontaminated natural sinks. If the economic activities keep growing, growth in pollution also continues and will start to work as a limiting factor to standards of living. At some point, the pollution will become so high that its negative impact outweighs the positive impact of economic growth on well-being. Figure 9 shows the system depiction of the *Limits to Growth* principle as it pertains to this example.

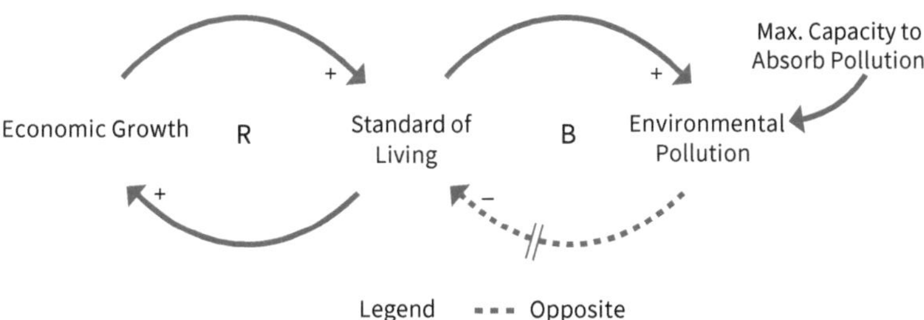

Figure 9. Systems depiction of the Limits to Growth principle. Source: Figure by author.

You could draw a similar figure for potable water, fossil reserves, or metals, amongst others. It could apply also to social, not just natural, capital. For example, another byproduct of economic growth has been rising income inequality. As detailed in Chapter 5, income inequality becomes a balancing feedback loop on welfare if it increases beyond the point where it starts to tear at our social fabric.

If these examples feel somehow less intuitive to you, that would be understandable. You and I have grown up in a world where perpetual economic growth was normalized, and we have come to expect it. We switch banks over a difference of a few percentage points in our savings rate, plan retirement by investing in the stock market, and organize neighborhood protests if local development plans threaten the increase in the annual value of our houses. However, from a historical perspective, sustained economic growth has been a recent achievement, as is illustrated by a plot of GDP over a longer horizon than usually depicted in Figure 10 (Maddison 2006).

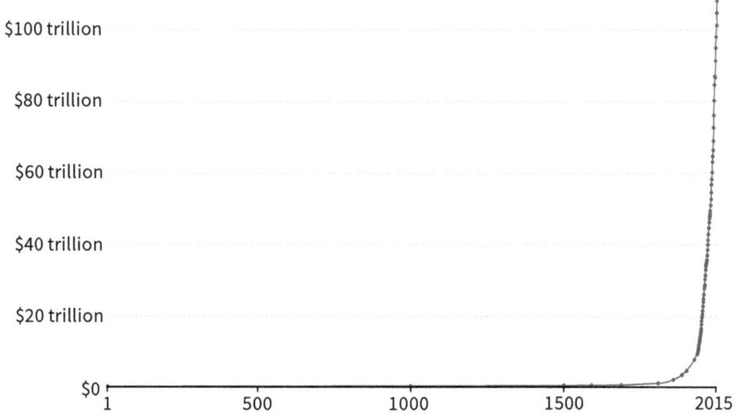

Figure 10. GDP over time. Source: Reproduced from Our World in Data (2022).

GDP figures from before 1500 are rough estimates, and this data series only has a few data points in that period. You should also keep in mind that exponential-like plots such as this one can look very different depending on the time scale used. Figure 10 only serves as an illustration of the qualitative work of historians who have documented that for much of human history; the idea of continued economic expansion was considered delusional, if not immoral (e.g., McCormick 2000). There have been warnings about the dangers that a quest for continuous growth can pose to society in more recent history too. One such warning came in the 1970s, from a team of scientists from MIT.

Several decades ago, a group of politicians, scientists, and artists called the Club of Rome (CoR) were contemplating what they called the "Continuous Critical Problems" (Meadows and Meadows 2007). Why, they wondered, do world problems like poverty, war, pollution, crime, oppression, resource depletion, terrorism, economic instability, racism, and drug addiction keep plaguing humanity? The CoR members had an intuition that they were interrelated. But how? They decided to task a small group of scientists to find an answer to this question with a new way of thinking and modeling. Together with the help of the aforementioned MIT professor Forrester, Donella Meadows, Dennis Meadows, Jørgen Randers, and William Behrens III used the first system dynamics model of the world, World3, to investigate the CoR's question. The general dynamics they found surprised them at first (Meadows et al. n.d.). The simulated world behavior was, although much more complex, not unlike what the *Limits to Growth* archetype would produce, and thus pointed to an answer that at the time, seemed anathema. Over the past five decades, this team published three books about their findings. The first book, called *The Limits to Growth* was published in 1972 (Meadows et al. 1972). The second book was published in 1992 (Meadows et al. 1992), and the third and last one in 2004 (Meadows et al. 2004).

As you might already have picked up, when referring exclusively to the first book, I will use its title *The Limits to Growth*, i.e., including "The" (still shortened to *LtG*). To all three I will refer as the *Limits to Growth* books, or *LtG* books. I will describe each book in more detail in the next section. But given that all three *LtG* books used scenario runs from World3, amongst other things, to support their message, let's dive into that model first.

2.2. The World3 Model

The World3 model consists of many interacting stocks, flows, and rates. Some examples of stocks are industrial capital, population, the total surface of arable land, and non-renewable natural resources. A few of the many flows are industrial output, deaths and births per year, and annual pollution generation. Industrial capital depreciation, fertility, mortality, and service capital investment rate are examples of rates. As with any SD model, the causal links between these variables are World3's key characteristics. It is those interactions that, as I put it in my thesis (Branderhorst 2020), "enable one to analyze global society as a system, i.e., as a world where the influence of policies and major trends are not always linearly proportional in impact, nor always felt and responded to immediately, and do not neatly stay within sector or country boundaries".

There are five interacting subsystems in World3: population, industrial output, food production, non-renewable natural resources, and pollution. Figure 11 shows a CLD of some of the interactions between these five subsystems.

The interactions in Figure 11 do not constitute the SD model. The full World3 model is more complex and contains many more variables. Appendix A contains an overview of all the variables and their interactions as modeled in World3 from the original book, and a more detailed and technical analysis of how the model behaves can be found in a publication by Dennis Meadows (1974). Because of its relative simplicity, however, the above CLD is more effective in giving people an idea of the general dynamics modeled in World3. Some works on World3 still mention the subsystems as the five *LtG* variables. However, the graphs of the 1972 *LtG* book depicted eight variables: population, fertility, mortality, industrial output per capita (p.c.), food p.c., services p.c., fraction of non-renewable resources remaining, and persistent pollution. Two more variables were added later in the 2004 update from the *LtG* team, as I will discuss in the next section on the three *LtG* books.

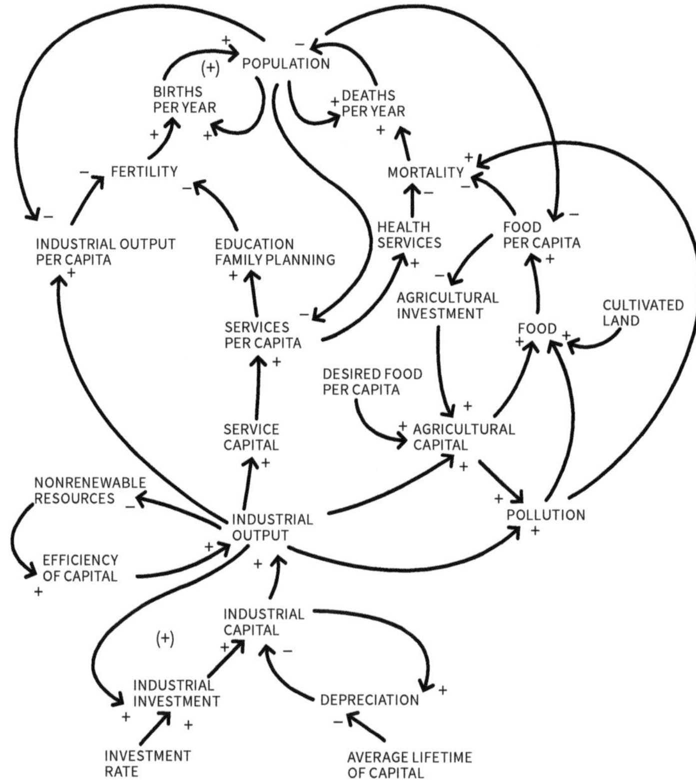

Figure 11. Causal loop diagram of World3. Source: Reprinted from Sverdrup and Ragnarsdóttir (2015).

2.3. LtG Publications and Scenarios

The *LtG* authors published three books over the course of 32 years. Each one contained an analysis of the global societal system based on World3 scenario runs. For each run, World3's parameters were set according to the specific scenario's assumptions. For example, the stock of non-renewable resources could be assumed to be the current estimates, or double those. Or the rate of technological innovation could be set to triple that of the historical average. By varying World3's parameters in this manner, the different ways that the global system could behave over time could be studied, and through that, understood a bit better. Each book consists of the findings and conclusions of the authors based on these studies and their general expertise. Although each book contained both new and rolling insights, the general behavior that World3 revealed remained the same: unless assumptions about societal priorities were drastically different than those of recent history, the model runs indicated a halt in industrial capital growth at some point in the 21st century, with some scenarios also showing a subsequent sharp decline (i.e., collapse).

The first book, *The Limits to Growth* (Meadows et al. 1972), was commissioned by the CoR and introduced World3 together with twelve scenarios. The first scenario was "business as usual" (BAU), which ran on historical averages without any additional assumptions on changes in humanity's behavior. In BAU, as I also explained in my journal article (Herrington 2020), "standards of living would at some point stop rising along with industrial growth once the accompanying depletion of non-renewable resources had started to render these a limiting factor in industrial and agricultural production. Continuation of standard economic operation without adapting to the constraint of growing resource scarcity would then require increasingly more industrial capital to be diverted towards extracting non-renewable resources. This would leave less for food production, citizen services, and industrial reinvestment, causing declines in these factors and, subsequently, in population". BAU ended up getting by far the most attention from both the critics and proponents of the *LtG* message (more on criticism in Chapter 5). But there were eleven other scenarios, some of which did not end in collapse. Two of those eleven were "comprehensive technology" (CT) and "stabilized world" (SW). CT assumed "a range of technological solutions, including reductions in pollution generation, increases in agricultural land yields, and resource efficiency improvements that are significantly above historical averages" (Herrington 2020). The SW scenario contained all these assumptions, but on top of the technological solutions, an assumed shift in global priorities was added (Meadows et al. 1972). In SW, humanity made a deliberate choice to limit industrial output and prioritize health and education services from a certain year onwards (I will sometimes also refer to these as "human services" or just "services"). SW was the only scenario in which declines were avoided entirely.

The second book, *Beyond the Limits*, was published in 1992 (Meadows et al.). World3 had been recalibrated with two decades of additional data, and new scenarios were run with this updated version. The authors also had an alarming new message. In the first book, the authors stressed that society had the opportunity to avoid any collapse pattern by shifting priorities in alignment with SW. Their analysis suggested that such a shift would have allowed a relatively smooth transition, in essence not unlike a logistic curve path as in Figure 8, towards a global equilibrium. However, in this second book, the *LtG* authors concluded that society had now transgressed above the Earth's carrying capacity. Humanity was already in overshoot.

The third and last book, *Limits to Growth: The 30-Year Update*, was published in 2004 (Meadows et al.). The last *LtG* book contained ten new scenarios. These scenarios were similar, but not identical, to those from the first two books in assumptions. The *LtG* authors used a revised World3 model for these new scenarios: World3-03. The model revisions, amongst other changes, included the incorporation of two new variables: society's global ecological footprint (EF) and human welfare. The assumptions about technological development went even further above historical rates than the technology scenarios in the previous two books, which made this CT scenario more optimistic compared with its 1972 version. In this optimistic CT version, the new technologies do help avoid collapse. There are still some declines, however, because technology costs become so high that not enough resources are left for agricultural production and human services. The 2004 book also contained a new scenario, "business as usual 2" (BAU2). BAU2 is also business as usual, but with double the natural resources. This scenario was added to address the criticism that natural resources, especially fossil reserves, turned out to be more abundant than the 1970s estimates indicated. More abundant resources in the World3 scenarios do not avoid collapse. Its onset is simply delayed, and its cause changed from resource scarcity to pollution. Because of the relative resource abundance, business as usual continues for even longer than in BAU, until a breaking point at which so much pollution has accumulated that agricultural output and population experience a steeper collapse than in any other published World3 scenario.

The aforementioned four scenarios, namely BAU, BAU2, CT, and SW, run with the latest World3 version, are shown in Figure 12. These are the scenarios that I used in my research. I chose these four because in this book I will argue that humanity's challenge today requires changes at the generative level of our world vision. In other words, our systemic problems demand we examine our common narrative. Each of these four scenarios represents a contemporary story we tell ourselves.

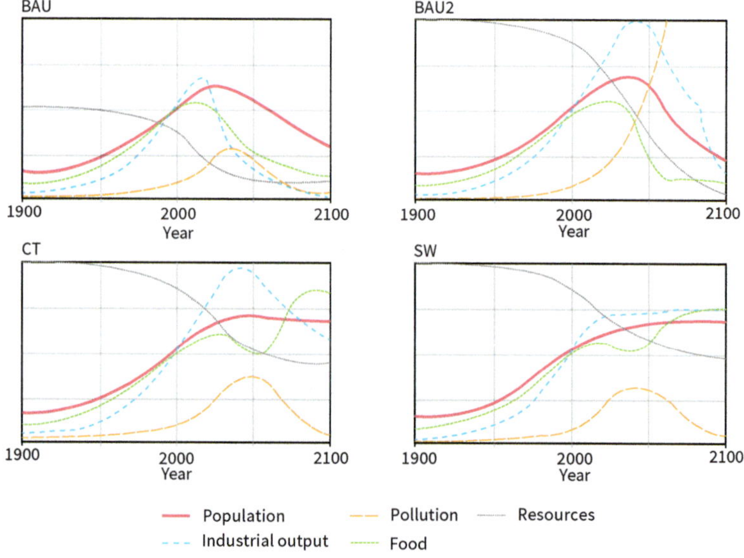

Figure 12. BAU, BAU2, CT, and SW scenarios of the 2004 LtG book. Source: Reproduced from Meadows et al. (2004).

BAU is the story in which we hang on to our "we can keep growing forever!" attitude that is tacitly ubiquitous in society until we can't anymore because we run out of resources. BAU2 tells the story in which humanity uses a relaxed constraint on resources to hold on to business as usual for even longer, which results in ecosystem breakdown from accumulated pollution, including greenhouse gasses. CT represents the technologist's belief that humanity can invent solutions for any environmental constraint and tells the story of a world that does not change its priorities much but avoids collapse anyway because of technology's salvation. In SW, humanity consciously lets go of expansionary growth as its ultimate pursuit. In this story, we shift societal priorities away from material consumption toward meeting human needs and protecting the environment, which avoids collapse and leaves us with the highest levels of well-being.

2.4. Updates to LtG

Several researchers have conducted qualitative reviews of *LtG* (e.g., Hall and Day 2009; Jackson and Weber 2016; Randers 2000; Saeed 2014; Sverdrup and Ragnarsdóttir 2014). These qualitative works focused on the general dynamics in World3, and each concluded that these dynamics accorded with real-world developments. For example, although the resource scarcity scare of the 1970s has somewhat subsided, many resources have in fact been getting scarcer, some to the point where it is threatening our ability to produce food and keep up industrial production (Bardi 2014), much like *LtG* forewarned. Unlikely ally Matthew Simmons (2000), an investment banker, former energy adviser to US President Bush, and a member of the National Petroleum Council, said about *LtG*: "The most amazing aspect of the book is how accurate many of the basic trend extrapolation[s] […] still are, some 30 years later" (p. 15). These qualitative reviews support the validity of the causal links in the World3 model. They also support the notion that the world is complex; that is, we are living in systems and thus should make sure to take their interconnectedness into account for good stewardship.

There was also one researcher who conducted a quantitative update on *LtG*: Graham Turner (2008, 2012, 2014). Turner collected global observed data for *LtG* variables and compared these with three of the twelve scenarios from the first book: BAU, CT, and SW. Based on this comparison, Turner concluded that real-world data aligned most closely to key features of BAU each time. After reading his work in 2019, I wondered whether this would still be the case based on the latest data. Additionally, I wondered if one would find the same results when using the latest version of World3, the one from the 2004 *LtG* book. Because Turner had used the 1972 variables and scenarios, his comparisons did not include BAU2, nor did they involve the two variables that were added in 2004, human welfare and EF. I was curious whether including the welfare and EF variables in a comparative study would change his conclusions. Lastly, given that BAU2's pollution crisis can be interpreted as depicting climate change (i.e., collapse from greenhouse gas pollution), I wanted to see how this scenario would fare in a data comparison against the other scenarios studied by Turner. One of the *LtG* authors, Jørgen Randers, performed a qualitative update in 2000 in which he postulated that not resource scarcity but pollution, especially from greenhouse gases, would cause this halt in growth. Randers concluded this after finding that non-renewable resources, particularly fossil fuels, had turned out to be more plentiful than assumed in the 1972 BAU scenario. This BAU2 scenario was also quantitively assessed in a 2015 recalibration study of World3-03 (Pasqualino et al.). The researchers found that society had invested more to increase food productivity, abate pollution, and invest in human services compared with BAU2. However, this did not necessarily mean that humanity had done so to the extent where collapse is avoided in World3. Pasqualino et al. did not compare their calibration with SW, nor did they use their recalibrated version of World3 to run the scenario beyond the present. Therefore, as the researchers themselves also pointed out in their paper, the study did not give any indication of whether humanity had increased pollution abatement, services, and food productivity rates sufficiently. Against the background of all this research, it seemed a pertinent question whether the empirical data available in 2022 aligned with BAU2, next to the other scenarios. There was no such quantitative update. So, I conducted the exercise myself.

3. Pursuing Growth Is the Cause, Not the Solution

This chapter starts with a description of my methods, which will be the same as in my journal article (Herrington 2020). The formulas can be skipped without consequences for your ability to follow this book's main thread. I will discuss my research outcomes, which sometimes differ from those of my 2020 article because of the updated data series. Based on the results, I will discuss the second insight we can derive from *LtG*.

3.1. What I Did

I examined to what extent the data available in early 2022 aligned with the recalibrated World3-03 (henceforth called "World3") scenarios. I compiled data from various official databases as indicators for what the following ten variables represented: population, fertility (birth rate), mortality (death rate), industrial output per capita (p.c.), food p.c., services p.c., non-renewable resources, persistent pollution, human welfare, and EF. I plotted these data along four World3 scenarios: BAU, BAU2, CT, and SW. As mentioned, these four scenarios form a comprehensive set of stories. The assumptions underlying each scenario span a range of technological, social, and resource conditions. The cause of decline, varying from a temporary dip to societal collapse, also differs for each scenario (Table 1).

Table 1. Description and cause of halt in growth and/or decline per scenario.

Scenario	Description	Cause
BAU	No behavioral assumptions added to historical averages.	Collapse due to natural resource depletion.
BAU2	BAU + double the natural resources.	Collapse due to pollution (climate change approximate).
CT	BAU2 + exceptionally high technological development and adoption rates.	Rising costs for technology eventually cause declines, but no collapse.
SW	CT + changes in societal values and priorities.	Population stabilizes in the 21st century, as does human welfare on a high level.

Source: Table by author.

3.1.1. Scenario Data

BAU, BAU2, CT, and SW, correspond to scenarios 1, 2, 6, and 9 in the 2004 *LtG* book. This means that, for the SW scenario, I assumed policy changes starting in 2002. To create the scenarios, I used the original CD-ROM that came with the 2004 book (I obtained a mint condition copy with the CD-ROM still attached). The CD-ROM contains simulations of the scenarios and numerical output of the variables. A zip file of World3-03 is also available from MetaSD (2022), and it can be run on free software from Vensim (2022).

3.1.2. *Determination* of Accuracy

To quantify how closely the *LtG* scenarios compared with the observed data, I used the same two measures as in Turner (2008):

(1) The combination of the following differences:

    a. The value difference (between the model output and empirical data),
    b. The difference (between the model output and empirical data) in the rate of change (ROC).

Both were applied at the time point of the most recent empirical data;

(2) The normalized root-mean-square difference (NRMSD).

The calculations of the two measures were carried out for 5-year intervals ending in the final year of the data series. In the below equations, that ending year is assumed to be 2020 to make the formulas easier to interpret. It is straightforward to adjust the equations for data series ending in another year.

3.1.2.1. Measure 1: Value and Rate of Change Differences

The difference in value was calculated using the following formula:

$$\Delta Value = \frac{(Variable_{2020} - (ObservedData_{2020})}{ObservedData_{2020}}$$

The difference in rate of change was calculated using the following formula:

$$\Delta RateOfChange = \frac{(Variable_{2020} - Variable_{2015}) - (ObservedData_{2020} - ObservedData_{2015})}{ObservedData_{2020} - ObservedData_{2015}}$$

3.1.2.2. Measure 2: NRMSD

In the formula below, the start of the sum is assumed to be 1990. This is what I used for each variable where this was possible; however, some series did not go back as far, in which case the below equation would have to be adapted accordingly.

$$NRMSD_{2020} = \frac{\sqrt{\frac{\sum_{t=0}^{6} \left(Variable_{1990+5t} - ObservedData_{1990+5t}\right)^2}{7}}}{\left(\frac{\sum_{t=0}^{6} ObservedData_{1990+5t}}{7}\right)}$$

These two measures do not provide the level of precision of some statistical tests, and they are not meant to. Precision does not always correspond to accuracy. The precision of linear regression and other econometric methods is based on assumptions of constancy in factors such as proportion of response, variance, correlation, and/or error distribution. These conditions can be assumed in controlled experiments or perhaps even in some very simple closed systems, but not in complex systems that involve human behavior (Forrester 1971; Meadows 2012). The measures I used are more appropriate for assessing the accuracy of the scenarios, given that they pertain to global developments. These measures should still be combined with a visual inspection, which is why a graph is presented for each variable in the next chapter. The accuracy measures complement the graphs by quantifying the alignment error. To evaluate this alignment error, it was necessary to choose uncertainty ranges beforehand. The uncertainty ranges indicate a level of close alignment between model variables and empirical data. Turner had chosen the uncertainty ranges of 20%, 50%, and 20% for the value difference, ROC, and NRMSD, respectively. I maintained these uncertainty ranges, firstly, for consistency with Turner's earlier *LtG* research work, and secondly, because I deemed the ranges suitable. They are wide enough to accommodate the low precision of World3 and the error margins we can expect in the empirical global data (specific measurement difficulties for each data source are discussed in the next chapter). At the same time, the uncertainty ranges are still narrow enough to be a meaningful indication of agreement between the observed and simulated data.

So, what did these accuracy measures indicate about World3's alignment with empirical data?

3.2. What I Found

The below table and graph provide an overview of the two accuracy measures for each variable and scenario (the graphs for each variable will be discussed in the next chapter). Table 2 shows the results for accuracy measure 1, and the graph in Figure 13 shows accuracy measure 2. As discussed in detail in Chapter 4, some variables had more than one data proxy. This means there was more than one data series against which to compare the scenario. In these cases, accuracy measures were calculated for each proxy. These are listed in the same variable cell in the table and separately displayed in the graph. For most variables, it was necessary to scale the data series, because the World3 output is given in different units than the empirical data. The variables that were normalized to the *LtG* value are indicated with an asterisk in Table 2, and the scaling factors for each variable are described in the next chapter.

Table 2. Accuracy measure 1: value difference and rate-of-change difference (in %). The variables that were normalized are indicated with an asterisk (*). Outputs are for or around the 2020 data point.

Scenario		Population	Fertility	Mortality	Food p.c. *	Industrial Output p.c. *	Services p.c. *	Pollution *	Natural cap.p.c. *	Wel Fare *	EF *
BAU	Δ value	−6	−17	18	−20	−20;−22	−5;−11	−20; 69	−10;−13; −4;−17;15	−5	15
	ΔROC	−42	20	702	−345	−161; −121	−21; −8	−14; 178	39; 130; 62; 190; 16	−151	0
BAU2	Δ value	−5	−10	8	−18	−7;−10	−11;16	−20; 69	−10;−13; −4;−17;15	−2	27
	ΔROC	−28	30	265	−245	−13; −48	−185;80	−14; 187	39; 130; 62; 190; 16	−71	426
CT	Δ value	−4	−10	6	−15	−7;−10	−11;16	−20; 69	−10;−13; −4;−6;16	−1	22
	ΔROC	−25	29	279	−178	−13; −47	−185;80	−14; 181	31; 115; 51; 171; 9	−52	201
SW	Δ value	−10	−22	12	−14	−14; −21	−18; 24	−19; 72	−9;−12; −3;−16;17	−1	11
	ΔROC	−52	−67	348	−256	−100; −103	86; 28	−8; 195	17; 90;34; 140; −4	−72	−75

Numbers within uncertainty ranges in green; outside the ranges in red, and on the uncertainty boundaries in black. Source: Table by author.

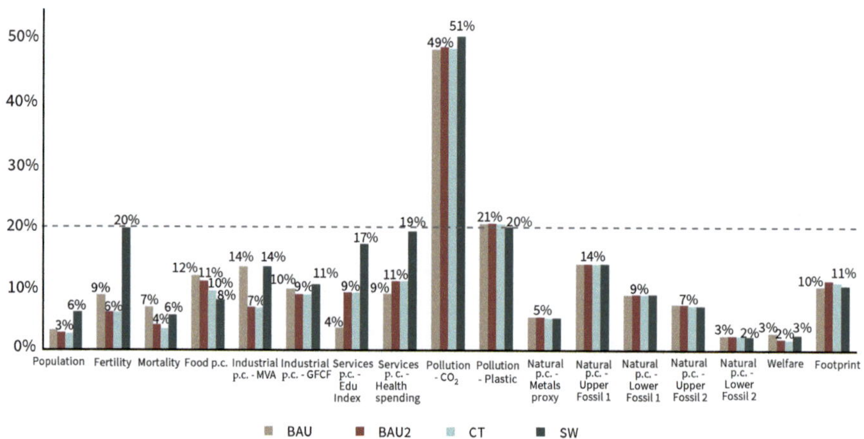

Figure 13. Accuracy measure 2: NRMSD. Plotted for each scenario and variable. Outputs are for or around the 2020 data point. Source: Figure by author.

The numbers in Table 2 that were within the uncertainty ranges (20% for the value difference and 50% for the ROC) are printed in green, and the ones outside the range are shown in red. The uncertainty boundaries were left in black. The 20% line is easily identified in Figure 13.

3.3. World3 Accuracy

In Chapter 4, I will discuss the fit for each variable separately. The discussion includes data sources, data reliability, and proxy construction, and is accompanied by a plot with empirical data and the four scenarios for each variable. But I started with an overview here because it gives a good impression of the overall alignment of the scenarios with the observed data.

When it comes to value, both measures indicate an overall close fit. Table 2 shows that most differences in value were also within the 20% range, except for pollution, the EF in CT and BAU2, and fertility in SW. The ROC showed more and larger deviations between the scenarios and empirical data. Measure 2 (the NRMSD) was not greater than 20% for all variables (Figure 13), except for pollution.

A close alignment to empirical data is not the only evaluation criterion for an SD model. More important is the validity of the causal links, as well as the structure of the model (Barlas 1996). Any model is subject to criticism, and World3 is no exception (*LtG*'s criticism is discussed in Chapter 5). However, the World3 structure is based on valid, observable global interactions, as the earlier qualitative *LtG* updates have shown. Its causal links are supported by sufficient confirmatory research where, together with the overall close alignment in value between the *LtG* scenarios and empirical data, it constitutes a firm testament to the work of the *LtG* team. World3's solid foundation and its alignment with the empirical data today also mean that the general dynamics described in the model should be taken seriously because when social, environmental, and economic factors interact, systems behavior will occur, including systems archetypes like *Limits to Growth*. And in that light, it is noteworthy that the scenarios show a halt in growth within a decade or so from now.

3.4. *The End of the Growth Pursuit*

The *LtG* team had been tasked with finding an answer to how the "Continuous Critical Problems" interrelated, and why they kept recurring despite humans' efforts to end them. After studying World3, they believed they found the answer. In a memoir, Meadows and Meadows (2007) recall how they reported back to the CoR:

"There is a primary cause of the Continuous Critical Problems: It is growth. Exponential growth of energy use, material flows, and population against the earth's physical limits. That which all the world sees as the solution to its problems is in fact a cause of those problems."

The World3 model signals that continuing business as usual, i.e., pursuing continuous growth, is not possible. Even when paired with unprecedented technological development and adoption (as in CT), business as usual would inevitably lead to declines in industrial capital, agricultural output, and welfare levels within this century. According to the World3 dynamics then, our choice is to either choose our own limits or have them forced upon us. Although economic growth is not an explicit variable in World3, it is spurred by industrial, agricultural, and population growth. A steep decline in those variables would inevitably also lead to an economic collapse. The findings of this research put the recent, relatively low economic outlook predictions and talks from organizations like the World Bank (WB 2022a) and the International Monetary Fund (IMF 2022) about a "pronounced slowdown" (Lawder 2019) of global growth in perspective.

3.4.1. "But How Will I Pay the Rent?"

Maybe you are wondering whether there will be an economy at all. If we don't focus on income or profits, how will we stay employed or in business? How will we eradicate poverty? After all, the promise of the growth pursuit is that, eventually, material wealth will be delivered to anyone willing to work hard. This promise justifies poverty in the now, even next to enormous wealth at the same time, because supposedly, the rich just got there a bit quicker than the rest. But the assumption that we need growth to have an economy, and that without having it as the primary goal we will inevitably spiral down into poverty or even anarchy, is worth a re-examination.

Firstly, it is not as if the world today is free of all forms of poverty and conflict, despite a relentless quest for growth over the past few decades. Plenty of people lost their homes during the GFC or the COVID-19 pandemic, even in rich countries. In fact, especially in these advanced economies, due to dynamics such as the ones I will describe in Chapter 5, the poor have only gotten poorer over the past few decades (Chancel et al. 2021). Many of the richest countries in the world are members of the Organisation for Economic Co-operation and Development (OECD 2018), and even in those countries, on average, almost one out of seven children still live below the poverty rate. If the pursuit of growth was the solution, would the past few decades not have been enough to eradicate poverty at least there?

Secondly, it is not self-evident that ending poverty could not also be achieved by means other than pursuing growth. Yes, over one billion people worldwide have been lifted out of poverty since the 1990s (WB 2018), most of them in Asia. This is progress, and we need more of it. But that is just it: It does not seem this growth-pursuit-spurred reduction in poverty can get us to a complete eradication. Let's leave aside the fact that poverty levels have seen a slowdown in decline over the past few years, and even increased again in 2020 (WB 2020). The reduction in poverty over the last few decades has come at high environmental and social costs, some of which are increased local pollution, massive wildlife extinction, soil degradation, and climate change. Based on the average global biocapacity, we need more than two planet Earths if everyone were to live at the level of the average Chinese, let alone the average Australian, which would require almost five planet Earths (GFN 2021a). Development economics is a broad field, and any meaningful argument should include regional specifics. In general, however, the assumption that the only way to eradicate poverty is to put a growth pursuit at the center of economic development seems hastily made. GDP is a national income flow. But each world nation also has wealth, vastly differing amounts of it. Could another way of reducing poverty be to share current wealth more equally, instead of relying on newly generated national income? Wealth sharing does not require growth and can be achieved with a much smaller ecological footprint. However, sharing more equally instead of pursuing growth might mean that the wealthy would have to satisfy themselves with less. Indeed, growth does seem the only way to lift people out of poverty without the elite having to give up anything. Part of re-examining the assumption that economic growth is the only way to battle poverty and keep a roof over our heads is to ask: who benefits from that story?

That said, advocating for a different goal than growth is not the same as being anti-growth. I might have reported back to the CoR that not growth itself but the pursuit of growth is the root cause of our perpetual problems. In developing economies, although prosperity could still be more accurately approximated with a different measure, even growth narrowly defined by GDP makes sense. And the necessary shift from fossil fuels to renewable energy in all countries is bound to spur growth in the solar and wind sectors. Striving for a different purpose than GDP growth or profits is not the same as advocating for de-growth. De-growth has received some criticism from scholars, for one because it doesn't seem very effective in practice (e.g., van den Bergh 2011). This makes sense in light of my research's finding that we need a shift in focus away from growth. Going in the opposite direction of what we were doing before is not that mindset shift; being anti-growth still is a focus on growth. To give a simple example, let's say I focus too much on eating, have become obese, and feel I should now lose weight (I am not saying anyone should lose weight to adhere to a certain standard of beauty, but in this hypothetical example, I do want to lose weight). I can go on a diet, which means I am still focusing on food, only in a different way. I may succeed in trying to lose weight, although we all know that diets do not work nearly as often as people try them. But if losing weight would be beneficial for me, the same benefit can be achieved through a more ambitious goal of focusing on my overall health. That focus would also yield a lot of additional benefits, like more energy, perhaps. It would also avoid the risk of going too far in the opposite direction and becoming too thin, which could be even worse for my health. Similarly, a society that centers around a different goal than growth can still be innovative and deliver well-being. Depending on that new goal, for which I will offer a suggestion in the next chapter, it might even be more innovative, deliver higher well-being, and have some other benefits on top of that. Sometimes, its economy might even still be growing. We just won't be preoccupied with it anymore.

3.4.2. Maybe We Just Need to Get Rid of GDP, Though?

If your mind works in any way like mine, you might now be thinking: "What if we just measure growth differently? Surely aiming for ever-continuing development, personal growth, or happiness would be a lofty goal?" Yes, the growth mindset is pervasive. But there is a problem with striving for perpetual personal development or ever-increasing citizen satisfaction. This issue is not that these things are hard to measure. They are not straightforward to quantify, of course, but neither is the economy. GDP is not even one of the best measures we could be using. It excludes many things that are vital to an economy, care work being the prime example. If the market is the engine of the economy, then cooking, cleaning, and caretaking form the hidden chassis of the economic car. The economy and indeed society would fall apart without these "invisible hands" (Monbiot 2017) doing the care work, of which the global economic value is estimated to be USD 11 trillion a year (Berkhout et al. 2021), but it is not measured in GDP. On the other hand, GDP counts things that we could easily live without, such as plastic-wrapped pre-peeled bananas, jarred farts from a *90 Day Fiancé* celebrity, tooth diamonds for a sparkling smile, and "This Smells Like My Vagina" candles. GDP can also easily be boosted by actions that damage longer-term economic growth potential, such as mass deforestation and overfishing (WB 2021a). GDP is good with quantity but not quality. Inflated health care costs in the US increase GDP without necessarily translating into any health improvements, while increases in health status might not boost GDP at all, or even bring it down because of the avoided future medical interventions (Pilling 2018). GDP is not great with any avoided costs because it only measures what is priced in the market. For example, avoided costs from climate-positive corporate leadership and governing do not easily translate into GDP, but climate-change-induced hurricanes boost local GDP through rebuilding efforts. Lastly, even in its narrow definition, GDP doesn't measure national income benefits for its citizens well for the very simple reason that the most used figure is total GDP, not GDP per capita. This means that GDP can increase even when per capita GDP goes down, as long as the population growth rate is higher than the rate at which GDP per capita is shrinking. Yet, as imperfect a measure as a country's GDP is, global society still mostly uses GDP to track our growth anyway.

I would argue that happiness (or "life satisfaction", these terms are used interchangeably in this book) is the best measure for society. In theory, what better purpose for a society could there be than to increase its citizens' happiness? It is entirely possible to construct a measure for happiness using surveys, possibly supplemented with other data. Bhutan has been governed by a "Gross National Happiness Index" since 1972 (coincidentally, the same year that *LtG* came out). For a decade now, the World Happiness Report, with its famous country rankings by average life satisfaction, has served as a foundational text for the annual United Nations (UN) High-Level Meetings. You have personal experience with measuring satisfaction too, as you have most likely submitted your satisfaction rating more than once for all kinds of products and services, prompted by a message asking you how happy you were with them. The resulting aggregate number, typically a satisfaction rating between 1 and 5 stars, is used by consumers and producers alike as a strong indicator of perceived quality. But that fifth star is the point; you cannot go beyond that maximum. That is the problem with wanting to grow your people's happiness levels, the same problem of pursuing continuous economic growth: Ever-lasting growth on a finite body is not possible. As discussed in Chapter 2, growth of any kind will, at some point, always encounter some limiting force. This holds true for our own bodies just as much as it does for the body of the Earth. We normally do not stay at high (or indeed very low) levels of happiness (Lustig 2018), unless we have a health condition, such as clinical depression. On a scale of 1 to 10, our body's biochemistry over time just reverts to a state that we rate around a 7, no matter how we started out. Like you, I suspect, I strive for personal growth. But I don't see this journey as an ever-upward trending line. Apart from the obvious fact that one day I will die, at which point of course any growth journey is definitively over, even in the entire middle-age phase, I see my personal growth more as a dynamic equilibrium. Sure, I have made steady progress over the past decade in some areas, but I have been distinctly declining in others. For example, my cognitive capabilities have grown, most notably perhaps my ability to fathom complex problems. But I have also noticed a steady increase in the amount of time I spend each week standing in a room thinking "what did I come in here to do again?" According to my mom, a wise woman who still reads a book per week at the age of 68, I won't be bothered by that for much longer because soon I will lose the ability to keep track of that weekly time.

I did not realize all of this at first, even after reading the *LtG* books. When I was writing my thesis on *The Limits to Growth* for my master's degree in Sustainability at Harvard, my thesis advisor advised me to speak with Professor Sterman, who taught at MIT. About 15 minutes into the conversation with Professor Sterman, I made the argument to him that progress and prosperity should just become defined in a more enlightened way. I even mentioned Bhutan. He replied that, actually, he had just come back from a study visit there. He explained that the conclusion of his visit was that although a Happiness Index is a good measure to include in governing, it is still not good governance to try to keep it growing indefinitely. I was not convinced, so I replied with a non-impressive "Hm". I did my own research after our talk, ultimately leading me to the conclusions above which are, unsurprisingly, the very same Professor Sterman had pointed out to me. In my defense, Sterman is the director at the MIT System Dynamics Group and co-faculty at the New England Complex Systems Institute, has been practicing system dynamics for about as long as I have been alive, and was a student of Donella Meadows. From our contact afterward, I don't think he gave it a second thought. Still, I wish I had done my research before going into that first conversation with him. I am sharing this with you so that maybe one day it can save you from embarrassing yourself in front of an MIT professor!

That said, this kind of thinking is in the right direction. When we talk about satisfaction, development, or happiness, we are centering the discussion on humans and their needs. As it turns out, that is the third insight my research revealed.

4. A Better Goal: Fulfilling Human Needs with Respect for Nature

Despite all the scenarios showing a relatively close track, there were meaningful differences between them for some variables. Although these differences are not large yet, they provide important indications about the possible direction in which humankind is heading. This chapter starts with a look at each variable separately, along with a discussion of the extent to which World3 and the observed data aligned based on the accuracy measures. I will also discuss the data sources and their reliability. For those wishing to explore the data for themselves, an Excel file with the data I used was provided as supplemental information in the journal article (Herrington 2020), and the updated version used for the research in this book can be requested through MDPI publishing. Those more interested in the narrative of this research than the details of the data used in it may want to skim this first section and pick the thread back up at the next one. In Section 4.2, I will elaborate on what the combination of the variables' alignments indicates, followed by my concluding insight in this chapter's last section.

4.1. Closest and Least Close Alignments

I start by describing the general aspects of my approach for compiling the data, before diving into the comparison results per variable on the next page. The data series extended to between 2017 and 2020. The graphs are in 5-year intervals, which means that in some cases, the most recent data point is either not depicted or is plotted against an *LtG* scenario output that is a few years off. Specifically, for the one case where the data series extended only to 2017, the 2015 data point is plotted against the 2015 *LtG* scenario number. When the series extended to 2018 or 2019, which was the case for several variables, these data points were plotted against the 2020 *LtG* scenario number. All the accuracy measures were calculated using the most recently available data. For example, for the proxy where the data series extended to 2017, the accuracy measures were calculated with the 2017 figures, even though 2015 is the last empirical data point plotted in the graph. Sometimes, empirical data were expressed in different units than in the *LtG* scenarios. In these cases, I normalized the data series to the 1990 scenario value, because that is the year that World3 was recalibrated to last (Meadows et al. 1992).

Some variables required proxies because the variable in World3 is not directly observable or quantifiable in the real world. For example, the population variable required no proxy because these figures are relatively easily counted, and there is no disagreement about the unit (one person, one count). But the human services variable, for example, required a proxy because services are not as straightforwardly defined. For several variables, I used the same data sources as Turner for proxies. However, in several cases, I was also able to improve on his proxies thanks to new or recently enhanced indices and databases. For example, Turner used the literacy rate as a proxy for services in his work, which made sense at the time given the available global data on education (which were not that much). But by the time I conducted my research, the UN had constructed an Education Index, which is one of the proxies I ended up using.

4.1.1. Population

- Data Source

I used figures from the World Bank (WB 2021b), which in turn states as its sources: UN Population Division, census reports, and other statistical publications from national statistical offices, Eurostat, the UN Statistical Division, the US Census Bureau, and the Secretariat of the Pacific Community.

- Reliability

The global population will likely be one of the more accurate data series used in this research. According to the WB (2021b), census reports in some countries will be less frequent and/or of lower quality than in others, but variances in the data should be within the precision that I worked with. I checked the series against the population series from the UN too. The WB population data differ slightly from the UN figures, but the errors are around 0.5%, which is acceptable given the low precision of World3.

- Fit

As can be seen in Figure 14, the BAU2 and CT scenarios were the closest in alignment, whereas SW was the furthest off. The BAU scenario also still fell within the ranges I had set for the accuracy measures.

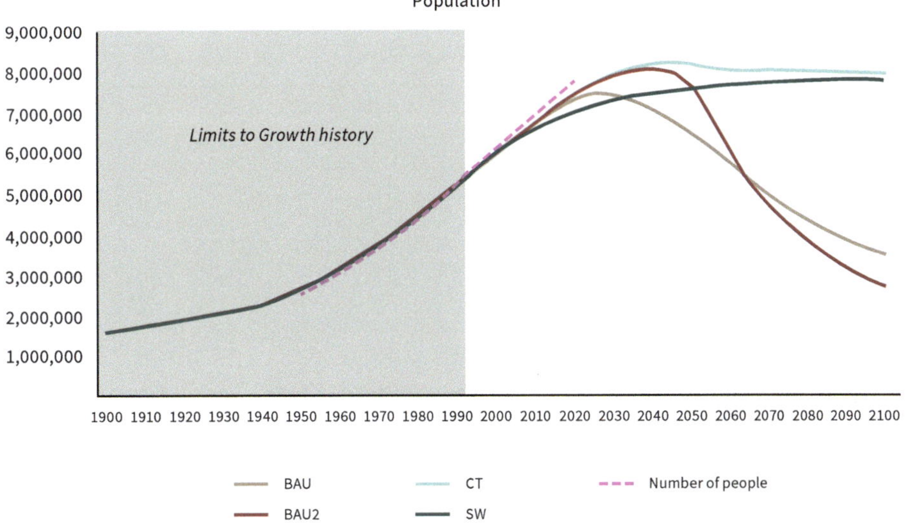

Figure 14. Scenarios and empirical data for population (in thousands of people). Source: Figure by author.

4.1.2. Fertility

- Data Source

I used the data series from the WB Open Data site (WB 2021c) for this variable. The WB mentions as its sources the same organizations and publications as the ones for its population series.

- Reliability

The reliability of this series should be as high as it is for the population series. Uncertainties around birth data can be higher in some developing countries. However, the WB (2021c) notes that its data "are generally considered reliable measures of fertility in the recent past".

- Fit

The birth rate was higher than in any scenario (Figure 15). The SW scenario fell outside of the uncertainty ranges for measure 1. The other three scenarios fell within uncertainty ranges for both measures, with BAU2 and CT aligning most closely with observed data.

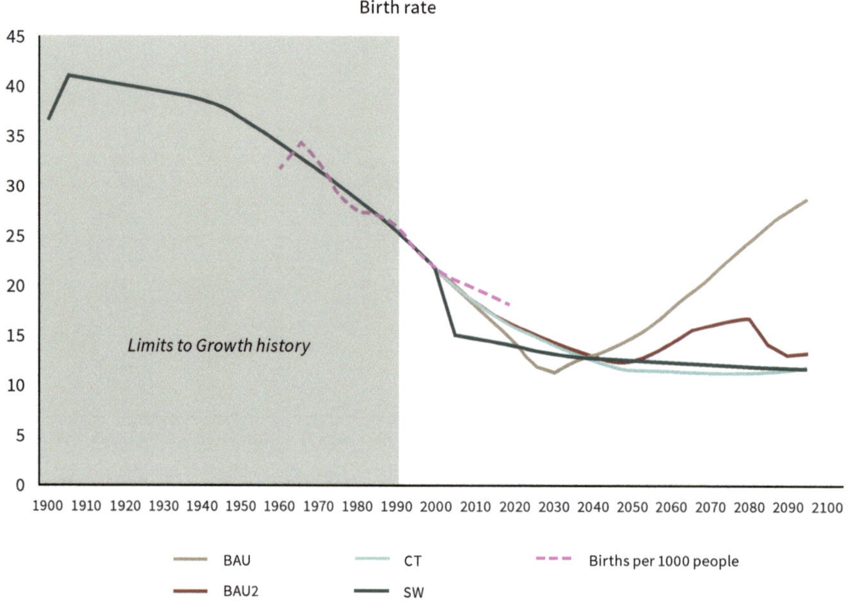

Figure 15. Scenarios and empirical data for fertility (births per thousand people). Source: Figure by author.

4.1.3. Mortality

- Data Source

I used the data series from the WB Open Data site (WB 2021d). The WB mentions as its sources the same organizations and publications as those for its population and fertility series.

- Reliability

As with the data in the previous two series, uncertainties around the data on deaths can be higher in some developing countries (WB 2021d), but this series' reliability should be similarly high as that for population and fertility and, thus, sufficient for this research. It should be noted that the death rate of 2020 can be considered an outlier because of the pandemic, during which mortality increased significantly. Instead of using the 2019 death rate, I kept the 2020 figure for two reasons. First, the 2021 death rate was not available at the time of this update, so we do not know to what extent the 2020 rate can be considered an outlier; it is possible global mortality maintains this higher level or indeed keeps increasing in the upcoming years (I am not saying I necessarily expect this but only that I don't know for certain one way or the other). Second, it could be argued that a virus outbreak like that of COVID-19 is part of the dynamics described in *LtG*. There is compelling research that climate change and human encroachment on wildlife habitats, among other factors, boost viruses (e.g., Gilbert 2022; Tollefson 2020), and climate change and habitat encroachment are direct results of our unsustainably high and growing levels of resource consumption and the pollution this consumption generates. Additionally, I checked the outcome using the 2019 mortality rate, and although this produces different numerical results, especially the ROCs, it changes nothing about which scenarios align most closely for any measure.

- Fit

As can be seen in Figure 16, all the scenarios aligned closely with the crude death rate in value and NRMSD, while the ROC was well out of bounds for each scenario. The CT and BAU2 scenarios showed a closer fit than the other two scenarios.

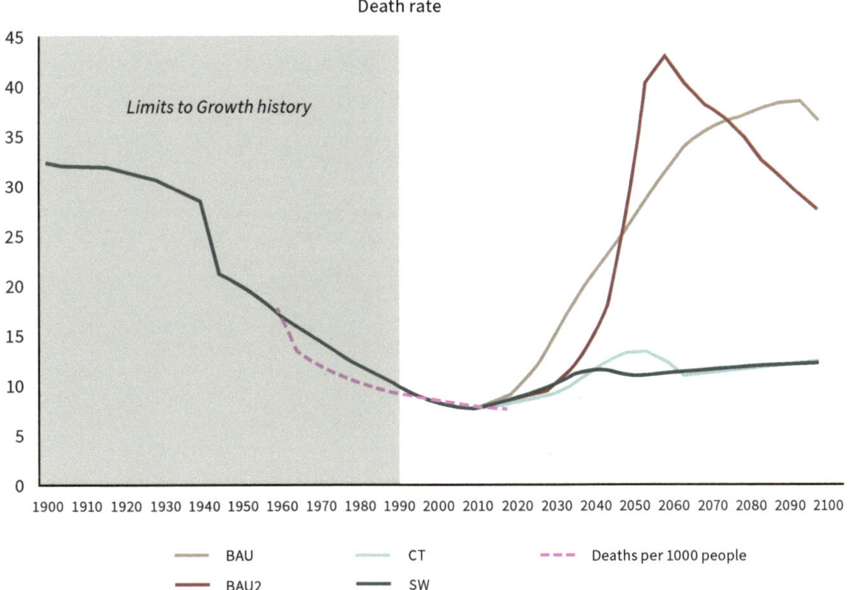

Figure 16. Scenarios and empirical data for mortality (deaths per thousand people). Source: Figure by author.

4.1.4. Food per Capita

- Data Source

I used the total energy available per person per day to approximate this variable. The daily caloric value per capita can be found in the Food Balance Sheets on FAOSTAT, the database of the Food and Agriculture Organization of the UN.

- Reliability

The FAO states that "there is a substantial amount of estimated or imputed data points", leading it to conclude that "the accuracy for certain products, countries and regions is not that good" (FAOSTAT 2021a). Because the FAO does not quantify the inaccuracy, I cannot say to what extent it impacted my research outcomes. FAOSTAT recently changed its methodology for calculating food balances, as evidenced by two separate databases: one is called "Food Balances (-2013, old methodology and population)" and the other "Food Balances (2010-)" (FAOSTAT 2021b, 2021c). Two years prior, I had downloaded the same data series for the previous comparison, which was not yet split up at that time. Therefore, I was able to compare some of the values from the old methodology with those from the new one. There were differences, but these were not material for the purpose of this research. For example, the old value for 2014 was 2887 kcal per capita per day, whereas this value on FAOSTAT is now 2906 kcal per capita per day, and for the year 2015, these numbers are 2898 and 2913, respectively. Because the change in methodology does not seem to have significantly impacted the numbers, given the margins of error I maintained, I chose to use the composite data series, meaning the data from 1960 to 2010 and then continued with the values under the new methodology for the years 2010 to 2019.

- Fit

Food p.c. was higher than in any other scenario, as can be seen in Figure 17. All scenarios compared favorably in NRMSD, with SW also being the closest in value. However, all scenarios were well outside of the 50% range when it came to the ROC.

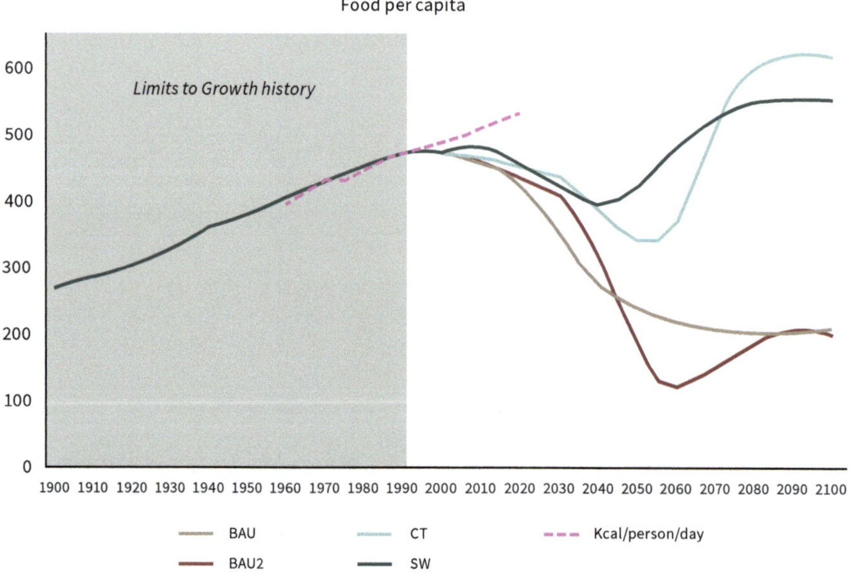

Figure 17. Scenarios and empirical data for food per person (in kilocalories per day). Source: Figure by author.

4.1.5. Industrial Output per Capita

- Data Source

The industrial output p.c. variable represented people's material and technological standard of living and was a factor in the ability of the World3 society to grow food and deliver services (Meadows et al. 2004). I used the manufacturing value added (MVA) and gross fixed capital formation (GFCF) as proxies. I divided both proxy series by population to arrive at per capita numbers.

MVA is a standardized macroeconomic indicator of an economy's real output in manufacturing (e.g., Moles and Terry 1997). Manufacturing refers to industries belonging to ISIC divisions 10–33, which include food, beverages, textiles, machinery and equipment, coke and refined petroleum products, basic metals, fabricated metal products, non-metallic mineral products, chemicals, paper and paper products, pharmaceuticals, computers and electronics, electrical equipment, vehicles and other transportation products, and furniture (UN DESA Statistical Division 2021a). Unlike GDP, MVA excludes retail and professional services, making it a suitable proxy for people's material standard of living. The MVA world series can be retrieved from the WB (2021e). It did not go farther back than 1997, so I standardized it to that year. The WB (2021f) also provides a global GFCF series, which I chose as a second proxy. GFCF includes land improvements (e.g., fences and drains), infrastructure (e.g., roads), building construction plants (e.g., schools, offices, hospitals, and industrial buildings), machinery, and equipment purchases. This also closely aligns with the definition of the industrial output variable in World3, especially as it relates to a society's ability to deliver services and grow food. There is some overlap between the two proxies, but they are also sufficiently different such that together, they complement each other well as proxies for the industrial output variable of World3.

- Reliability

The reliability of both proxies should be adequate for the purpose of this research. Given the mandate of the WB (2021f), one can assume they source from industry associations and government agencies. These are credible institutions, which, in turn, collect the data through regular censuses and firm surveys (Moles and Terry 1997). The WB (2021g) notes that data quality on fixed capital formation can be weak in some cases. However, I was comparing the trend in industrial capital growth for this variable rather than absolute numbers, because the *LtG* units differed from the ones used by the WB. This means that I normalized the data, which makes imprecision in the absolute numbers much less important than consistency in data collection. Given that there are no indications of material changes in the collection method of the WB for this proxy, the data series can be assumed to be accurate enough at the level of aggregation that I worked with. The GFCF series is based on the System of

National Accounts 1993 standards (WB 2021g), which ensures some standardization in reporting across national accounts (UN DESA Statistical Division 2021b).

- Fit

Figure 18 shows how the GFCF and MVA proxies gave similar results. Both series were closely comparable in value and NRMSD for all scenarios, but only BAU2 and CT also had an ROC within the uncertainty range.

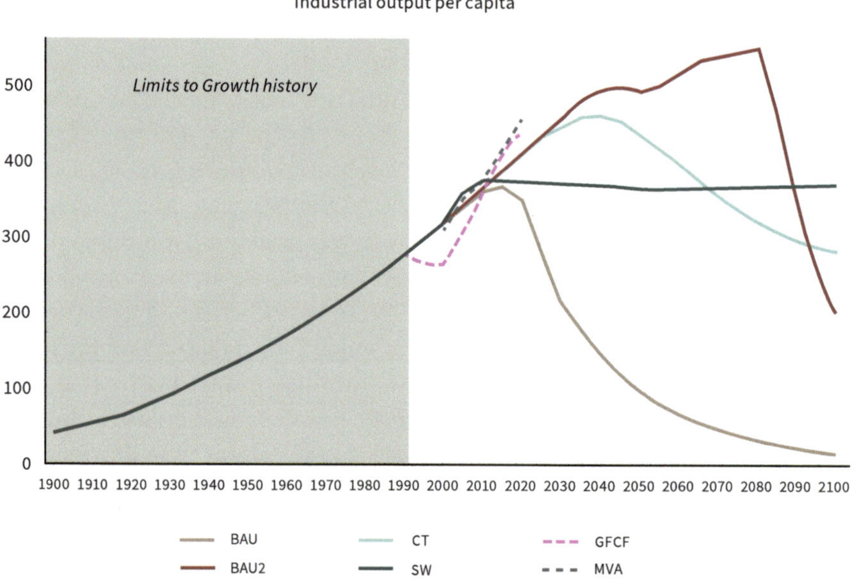

Figure 18. Scenarios and empirical data for industrial output (gross fixed capital formation and manufacturing value added). Source: Figure by author.

4.1.6. Services per Capita

- Data Sources

In World3, services p.c. represents education and health services (Meadows et al. 2004). I used the Education Index (EI) and national spending on health as proxies. The EI is constructed by the UN Development Programme (UNDP 2021a). It is calculated based on the expected years of schooling and the mean years of schooling. These two figures can differ, especially in developing countries, and thus combined, they provide a good indication of currently available education services. No such global index exists for health, but the WB (2021g) provides a world series for government spending on health expressed as a percentage of GDP. The *LtG* books each contain scenarios that end in collapse due to resources being diverted away from human services to industry so as to keep extracting natural resources, abate pollution, and/or produce food. The fraction of GDP is an indication of how society's resources are allocated, as expressed by the WB's statement that health financing is "critical for (…) people obtaining the quality health services they need" (WB 2021g). Tracking the fraction of global GDP spent on health can, therefore, help reveal whether the dynamic described in the books is observable in the real world.

- Reliability

The reliability of the EI proxy should be adequate for purpose of this research. The EI consists of census/survey information compiled by various official government agencies, which are widely considered reliable (Barro and Lee 2021; UNDP 2021b). UNDP only recently started to publish a global EI, so in an earlier phase of my research, I created one myself by weighing each country's EI by its population fraction. During this process, I discovered that the EI had missing data points, and for a handful of small countries, it did not have any data at all. However, I know from my own sensitivity analysis that leaving out those few countries and filling in incidental data gaps with the values from the first year the data become available (again) does not significantly impact the global EI.

Both GDP and health expenditure are widely and frequently recorded figures. The health spending series is sourced from the World Health Organization (WHO) and consists of "all health spending in a given country (…) regardless of the entity or institution that financed and managed that spending" (WB 2021g). The WHO (2021) collects data from "government budgets and health accounts studies", which should be sufficiently reliable given that my research does not require high-precision data. This is underlined by the WB comment that the series "generates consistent and comprehensive data on health spending (…), which in turn can contribute to evidence-based policy-making" (WB 2021g). The data series did not extend further back than 2000, so the series was normalized to that year.

- Fit

The proxies gave similar results, as shown in Figure 19. A close agreement in value and NRMSD was observed for all the scenarios except SW, but only BAU was also below 50% with regard to the ROC. Overall, the BAU scores best across both accuracy measures with this variable.

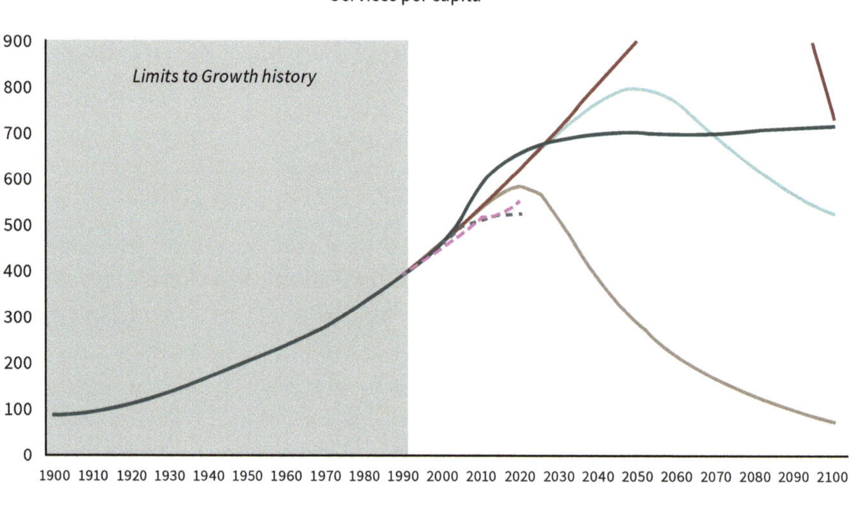

Figure 19. Scenarios and empirical data for services (health expenditure and education index). Source: Figure by author.

4.1.7. Pollution

- Data Source

World3 assumes pollution to be globally distributed, persistent, and damaging to human health and agricultural production. I used CO_2 concentrations and plastic production as proxies. The atmospheric CO_2 data (Dlugokencky and Tans 2021) were obtained from the National Oceanic and Atmospheric Administration (NOAA). I subtracted the 1900 CO_2 level of 297 parts per million (Etheridge et al. 1996) because the *LtG* scenarios put pollution at 0 in 1900. Although CO_2 is not the only persistent pollutant—NOx, SOx, heavy metals, and ozone-depleting substances are other examples— I chose it as a proxy because of the global impacts of climate change on human health, the environment, and our ability to grow food, and because accurate time series data exists for this pollutant.

The global plastic production data were sourced from Geyer et al. (2017). I adjusted the data downwards by the share of plastic that is discarded, which reportedly went from 100% in 1980 to 55% in 2015 (Geyer et al. 2017). Not all plastic is considered pollution; however, I considered it an appropriate proxy since plastic is persistent and ubiquitous in today's society. Various kinds of plastics can be found throughout the entire consumer product and food supply chain, from oceans and marine wildlife (van Sebille et al. 2015; Smillie 2017) to tap water (Kosuth et al. 2017), from agricultural land (Nizzetto et al. 2016) to dietary components and the air we breathe (Wright and Kelly 2017b), prompting a growing body of scientific literature on a wide range of possible negative human health effects (Halden 2010; Wright and Kelly 2017a). This led a coalition of environmental and human health organizations to conclude in a joint report that (Azoulay et al. 2019) "plastic threatens human health at a global scale".

- Reliability

The CO_2 data from credible organizations such as NOAA are widely considered reliable. NOAA (2021) uses air samples taken from remote sea level locations, which it claims, "results in a low-noise representation of the global trend". Their stated uncertainty is 0.10, or about 0.3%, and thus well below the uncertainty ranges I have maintained. The NOAA CO_2 data are very similar to global CO_2 averages published by other organizations that use different methods. NOAA also provides data series on persistent gasses of N_2O and SF_6. Adding these as CO_2 equivalents to the series did not make a significant difference in my analysis despite their higher global warming potential, because of their lower concentrations. To not unnecessarily complicate the method and data sources, I left them out of this research.

The models used to create the plastics data series contained multiple assumptions and simplifications, introducing considerable uncertainty in the estimates (Geyer et al. 2017). For this reason, the authors rounded cumulative results to the nearest 100 metric tons and conducted sensitivity analyses around mean product lifetimes and waste management rates. In these analyses, plastic estimates changed by a value between 4% and 8%. This is well below the 20% uncertainty range I used, so I assumed the plastics data to be accurate enough for this research.

- Fit

The scenarios had not started to diverge yet, so all showed the same comparison (Figure 20). Both accuracy measures were outside the uncertainty ranges for the CO_2 series. For the plastics proxy, measure 1 was within range for each scenario, whereas measure 2 was right on the uncertainty range and, therefore, inconclusive. The 2020 CO_2 data point could be considered an outlier due to the COVID-19 pandemic. The year 2020 saw a reduction in CO_2 emissions, but this lower level was not maintained when countries opened up again (Global Carbon Project 2021). This should be kept in mind with the slowdown in growth seen in this proxy. However, it seems unlikely that the absence of COVID-19 would have resulted in an alignment of this proxy within the uncertainty range. In my previous comparison, which used data from just before the pandemic, the CO_2 proxy was already out of bounds.

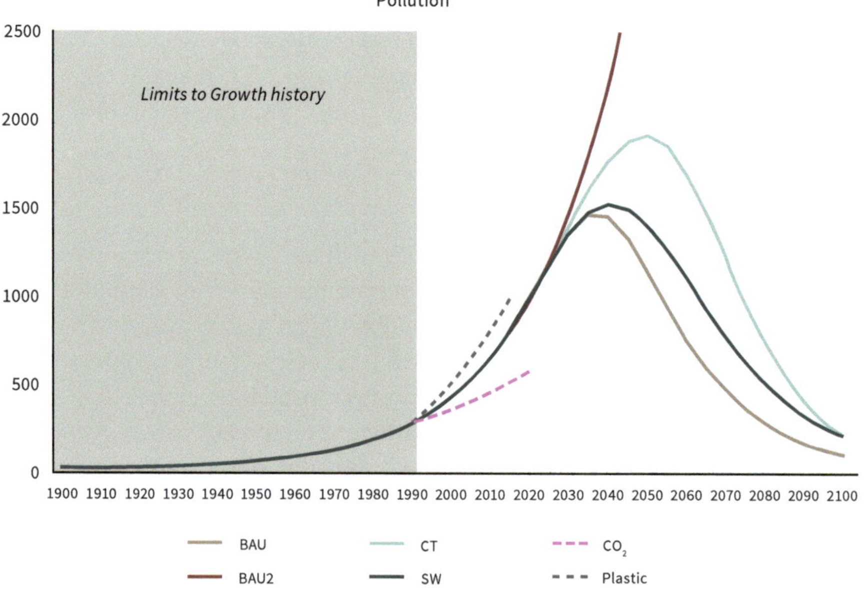

Figure 20. Scenarios and empirical data for pollution (plastic and CO_2). Source: Figure by author.

4.1.8. Non-Renewable Natural Resources (Natural Capital)

- Data Source

I used two fossil fuel proxies and one metal proxy. I assumed full substitution between energy or metal resources, which is conservative given the current state of technology (Brathwaite et al. 2010; Graedel et al. 2015; Henckens et al. 2016). The proxy data series that I created were not normalized to 1990 values because they represent fractions (i.e., they run on a scale from 1 to 0), so scaling them would distort the comparison. Because BAU and BAU2 differed only in resource amount and these were set to 1 at 1900, the two scenarios show the same curve.

Both fossil energy proxies consisted of estimates of remaining coal, oil, and natural gas (which would be named more accurately as "fossil gas"). The first fossil fuel proxy was the same as that in Turner's earlier work. Turner determined high and low expert estimates for fossil energy resources in 1900 from various expert reports and papers. He shared those estimates with me; Table 1 in his 2008 paper lists all the sources he used. The annual consumption of each resource was sourced from the World Watch Institute, which, in turn, had compiled the data from organizations including the UN, British Petroleum (BP), and the US Energy Information Administration. I updated Turner's series with the consumption data from BP's Statistical Review of World Energy (BP 2021) and summed over the values for the three fossil resources to arrive at the total annual production series. These production data were cumulatively subtracted from the total high and low resource estimates, resulting in an upper and lower bound for the fraction of non-renewable resources remaining over time. The second fossil energy proxy was constructed using the same method, but with resource estimates from a Geochemical Perspective (GP) publication (Sverdrup and Ragnarsdóttir 2014), and the production data from the WB (2021h). The WB, in turn, sources these production data from the International Energy Agency (IEA). The first fossil fuel proxy I used concerns the consumption data, and the second fossil proxy, the production data. So, not only do the two proxies differ in their starting point (total reserves) but also in their decline rates. This made them sufficiently relevant and complementary to each other, such that I decided to include them both.

The metals proxy consisted of resource estimates of 21 metals: aluminum, antimony, bismuth, chromium, cobalt, copper, gold, indium, iron, lead, lithium, manganese, nickel, niobium, palladium, platinum, silver, tantalum, tin, vanadium, and zinc. The resource estimates of the metals available in 1900 were based on the GP publication also used for the second fossil energy proxy (Sverdrup and Ragnarsdóttir 2014). The production data of each metal were obtained from the US Geological Survey (USGS 2021a, 2021b). GP provided the remaining recoverable amounts for each metal as of 2010, so I summed the USGS production values from 1900 to 2009 and added this sum to the metal resource GP estimate to arrive at the 1900 resource figure. The production and resource data were subsequently summed over the 21 metals, and the total annual production was subtracted from the 1900 total resources over time.

- Reliability

Although each fossil proxy was based on the data from credible organizations, non-renewable natural resource data were among the more uncertain compared with other variables in this research. It is simply unknown exactly how much of the resources are left in the ground because there is no direct way to measure them. Consequently, I worked with the upper and lower bounds of expert estimates for fossil fuels, which should mitigate the inherent uncertainty in fossil resource data sufficiently for a meaningful comparison. Turner (2008) deliberately created bounds for the fossil energy proxy that lay on the extreme ends of the spectrum. The high and low expert estimates from the GP publication for the second fossil energy proxy were closer together. I took some assurance from the fact that the second fossil fuel proxy fell between the upper and lower bounds of the first one. There were no upper and lower bounds for the metal proxy; however, it should be kept in mind that both the GP publication and the USGS numbers will come with considerable uncertainties. For some minerals and metals, the production data for 2018 were not yet available, so I used preliminary estimates from the latest USGS annual publication instead of their historical statistics. The production figures seemed to be somewhat different; however, the order of magnitude was lower than the uncertainty ranges I worked with.

- Fit

Because the scenarios had only started to diverge, all exhibited similar comparisons in value, as is also observable in Figure 21. All three proxies showed alignment errors below 20% with regard to value and NRMSD. The metals proxy showed a close alignment in their ROC as well. The lower bounds of the fossil proxies, for the most part, were also relatively close in terms of their ROC. However, the two upper bounds of both fossil proxies fell outside the range for the ROC.

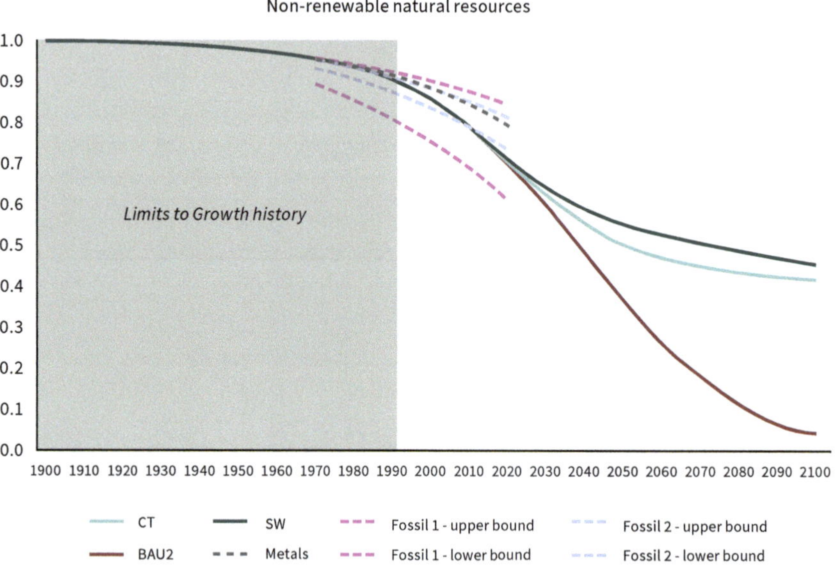

Figure 21. Scenarios and empirical data for non-renewable resources (one metal proxy and two fossil fuel proxies, both with high and low estimates). Source: Figure by author.

4.1.9. Human Welfare

- Data Source

This variable was created by the *LtG* team specifically to represent the UN Human Development Index (HDI). The HDI data series can be found on UNDP's (2021c) website. In the third *LtG* book, the authors (Meadows et al. 2004) note that the World3 welfare variable was very close to the value of UNDP in 1999. But this was no longer the case for the latest version of the HDI data series when I used it in my research. This is likely due to retroactive adjustments to the HDI series, resulting from methodological changes over the years (UNDP 2021d). The UNDP (2021d) states: "The difference between HDI values (…) published in HD Reports for different years represents a combined effect of data revision, change in methodology, and the real change in achievements in indicators". UNDP (2021d) therefore advises not to source HDI numbers from Reports, but to use the "data series available in the on-line database". Therefore, I scaled the current HDI data with a factor of 1.1 to line up with the value of World3 scenarios as of 1999.

- Reliability

The extent to which revisions to the HDI may have impacted this comparison beyond a scaling issue is unknown. The UNDP (2021b) states many sources for its HDI, including census/survey information compiled by various official government agencies, intergovernmental organizations, and academia. These are all considered credible sources and can be expected to produce figures sufficiently accurate for the purpose of this comparison.

- Fit

The HDI showed a close agreement in value and NRMSD for all scenarios, see Figure 22. No scenario was within range for the ROC, although at 53% CT was only marginally above it.

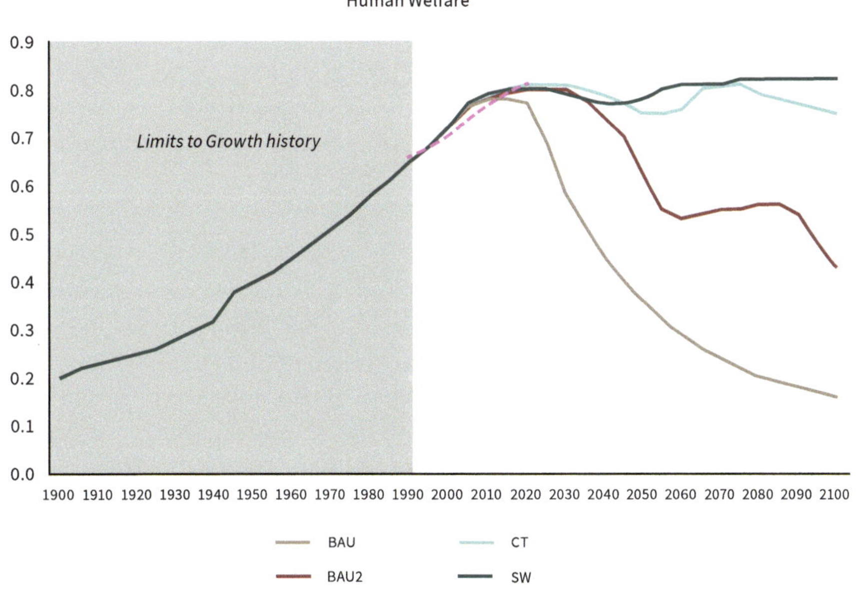

Figure 22. Scenarios and empirical data for welfare (UN Human Development Index). Source: Figure by author.

4.1.10. Ecological Footprint (EF)

- Data Source

This variable represents Mathis Wackernagel's ecological footprint (Meadows et al. 2004). The Global Footprint Network (GFN 2021a) publishes the EF on its website; however, its public dataset package only extended to 2018 at the time of writing. Lauren Hanscom, the GFN's CEO at the time, explained to me that this was because the GFN draws from UN sources as input data for the calculations. Consequently, in their public datasets, they are limited by the data years available from those sources, which often lag by several years. The GFN team provided me with their nowcasts—which they use to determine Earth Overshoot Day (EOD)—for the years 2018, 2019, 2020, and 2021. I compiled a composite data series from the public global footprint dataset supplemented with the nowcasts. Instead of the 2020 data point, I used the 2021 nowcast as the latest observation in this composite series. The year 2020 is an outlier because of the pandemic; it has by now been established that the dip in emissions and other demands on the planet were temporary, as CO_2 and other waste generation and resource use rebounded in 2021 (e.g., IEA 2022). EOD 2021 was July 29, the same date as 2019, the year before the pandemic (GFN 2021a). In the year of this book's publication, 2022, EOD fell on July 28, the earliest date in the history of EOD publications. I scaled this composite EF series to scenario values between 1990 and 2000 (with a factor of 1.16) because the *LtG* team would have calibrated World3 to line up with EF figures at the time. The reason that today's EF data did not exactly line up, Dr. Wackernagel (2021) told me in a personal email, is because of retroactive changes to the underlying data inputs. As mentioned, the GFN sources these inputs from UN entities, which periodically revise their methods, resulting in changes to the public datasets and thus the outcomes of the GFN's calculations.

- Reliability

The GFN (2021b) states that the "Ecological Footprint accounts provide a robust, aggregate estimate of human demand on the biosphere as compared to the biosphere's productive capacity". Although I don't think it's likely given the low precision of this research, revisions to the EF calculation from the underlying data changes may have impacted this comparison beyond what can be solved with scaling. The nowcasts for the last three years can be assumed to be accurate enough given the low precision I worked with. The fact that I used the nowcasts for the years 2018 to 2020 means that the methods used to calculate these last three years differ from the methods used for the other data points in the series, but this, too, should not be a material issue for the purpose of this research.

- Fit

Figure 23 shows the plot for this variable. Both BAU and SW were within the 20% bound for value, but only BAU was also inside the 50% range for the ROC. The EF was below 20% for all the scenarios relative to NRMSD. Overall, BAU aligned most closely with this variable.

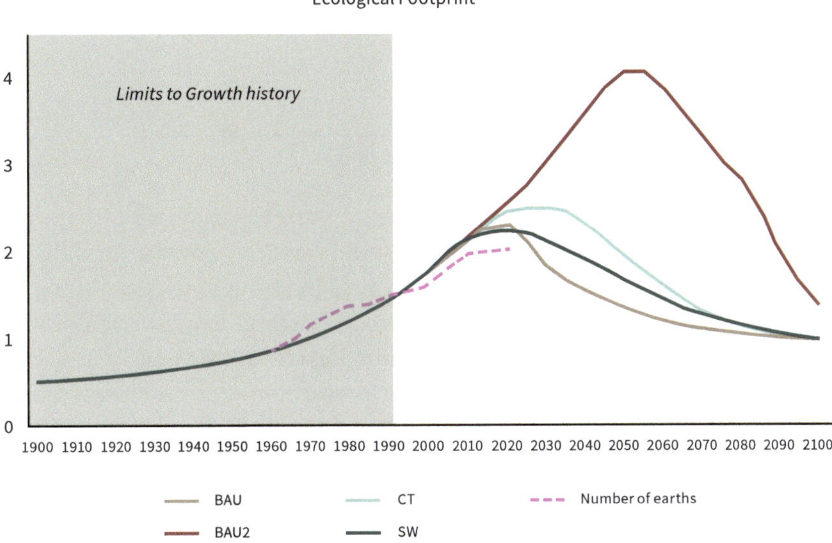

Figure 23. Scenarios and empirical data for the human ecological footprint. Source: Figure by author.

4.2. Most Closely Aligned Scenarios

As you can see from the above graphs, although the observed data relatively closely align with those of the scenarios at the moment, we can expect this to change in the near future because the scenarios start to diverge around the present time. Some do this later than others, which is why it was possible to distinguish the closest-aligning scenario for some variables but not yet all.

Table 3 contains a count per scenario for each time it was the closest fit. A scenario was counted as the closest fit when it aligned more closely than other scenarios, and at least one proxy was within the uncertainty bounds for at least one accuracy measure. When all scenarios were outside the uncertainty bounds for both measures, they were counted as inconclusive (the last column in Table 3). In one case, the food variable, I counted the scenarios as inconclusive even though they all aligned closely in NRMSD because the ROC was so much out of bounds. For cases where two or more scenarios aligned to the same extent, they were each counted. Choosing the closest fit was also not possible in the case where all scenarios aligned closely to a similar extent. This was the case with non-renewable resources, so all the scenarios were counted for this variable. This double or triple counting is why Table 3 shows 18 total counts over 10 variables.

Table 3. Count per scenario of closest agreement with empirical data.

Scenario	Count of Closest Alignment with Data
BAU	3
BAU2	6
CT	6
SW	1
None	2

Source: Table by author.

The use of more than one proxy for some variables did not lead to double counting; although different proxies for the same variable sometimes had different numerical results, they led to the same outcomes in terms of alignment (or not) to a certain scenario. In particular, the BAU2 and CT scenarios did not significantly deviate before 2020, resulting in both being the closest fits for several variables. These two scenarios aligned closest most often. The lowest count for the closest fit was for SW, the scenario that depicted a sustainable trajectory.

4.3. Stabilized World

Thus, my research indicates that global society is not on a sustainable path. However, remember that the scenarios have not diverged much yet. This means that it is not yet too late to make a directional change. But as you can see from the graphs, we will not reach this different trajectory by being a little less bad every year. Incremental change will not suffice for realignment with SW; for that, humanity needs to make major societal changes. And the most challenging part of it all would be the speed with which these changes need to be made. This sounds like a lot of effort, and it is. I could point out that dealing with the effects of ecological and societal collapse is also a lot of effort. But that would tell you nothing that climate and other scientists, including the *LtG* authors, have not already been saying better for many years. Instead, let's see what a world aligned with SW would look like.

As mentioned, SW has the same technological innovation rates as CT in pollution control, remediation, and regenerative technologies, but in addition, it consciously limits industrial output. It reprioritizes resources towards human services, i.e., access to education and health care. These services would encompass birth planning because the SW assumptions also include perfect birth control effectiveness. Additionally, the average desired number of children is assumed to be two per woman. The combination of the SW assumptions of prioritizing education and a lower desired number of children is in accordance with empirical research, which finds a strong negative correlation between a woman's education level and the average number of children per woman (e.g., Murtin 2013; Roser 2014). SW maintains the highest levels of welfare (Figure 24); in contrast to the first SW run in the 1970s, however, a temporary dip is already no longer avoidable in the 30-year update.

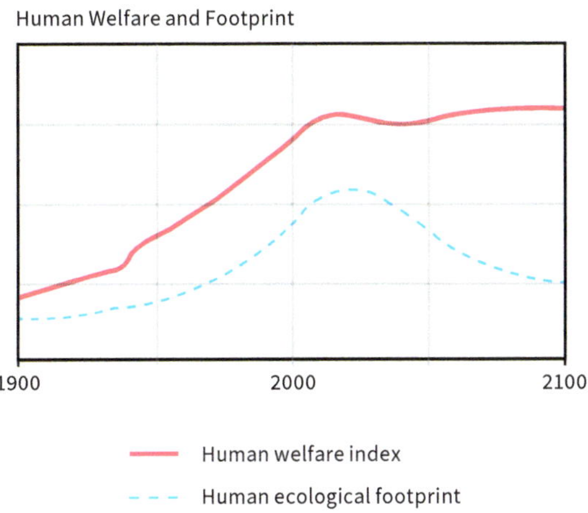

Figure 24. Welfare and EF developments for SW scenario in the third *LtG* book. Source: Adapted by Hilary Moore from Meadows et al. (2004).

A shift in focus away from industrial consumption towards health services, education, and pollution abatement does not just happen. Such societal changes coming about today would constitute a major and structural break from business as usual. The SW assumptions thus represent a fundamental change in our underlying values. But while pivotal, changing our values and priorities would hardly be a capitulation to grim necessity. On the contrary, imagine for a moment what a society that makes this change would look like. Prioritizing health and education over industrial output is accomplished by a society that centers itself around meeting human needs; not through growth but directly. A society that puts its resources towards cleaning up and preventing pollution is one that understands its interdependence within the ecosystem and realizes that respecting nature's limits is to respect oneself. The *LtG* graphs show how alignment with the SW scenario might lead to a more stable future, but they leave unseen how much more our natural surroundings would flourish and we would thrive. A world where human activity is regenerative instead of degenerative is not just one in which collapse is avoided; it is one bursting with life, zest, and wonder. SW realignment would constitute a drastic change in direction, but it would also be a drastic improvement. A lot of work, undeniably. But worth the effort.

5. The Sustainability Revolution: Humanity's Dying and Emerging New Narratives

Working towards some real-life alignment to the SW scenario requires we change almost everything about how we meet our needs today. Moreover, it requires we increase the extent to which needs are being met in the first place. We are also running out of time to make these changes, so we need to be fast. In short, calling for realignment with SW is calling for a sustainability revolution. Advocating for a revolution is not a call to take up arms. We have had many peaceful revolutions before, most notably the agricultural and industrial ones. It is, however, a call for defiance; an invitation to re-examine old mindsets and question existing power structures. This is what *LtG* already did 50 years ago—and, in my opinion, the reason why the message, and sometimes the authors personally, were attacked so viciously. This chapter starts with an overview of this criticism and my commentary on it. I will put this criticism against the background of dominant mindsets at the time and explore to what extent they can still be considered prevalent and useful, or not.

5.1. Criticism

The *LtG* books and World3 received much attention, some of which was positive, of course. A plethora of criticism spread with equal fervor, however (e.g., Norgard et al. 2010). Some economists critiqued World3's modeling assumptions, while others critiqued the modeling technique itself (system dynamics). Some of this criticism had validity; after all, no model is perfect, and thus no conclusion drawn from its output is indisputable. But a lot of criticism was unsubstantiated (Bardi 2011). At best, these criticisms could be classified as misinterpretations of the scenarios and key messages of the books. Yet, despite obviously being false, some of these misconceptions turned out to be the most persistent and influential in the public debate, so I will start the overview of *LtG* criticism with those.

5.1.1. Misinterpretations

One relatively well-known *LtG* criticism is the claim that the first book predicted resource depletion by 1990 (Passell et al. 1972). This misconception was promulgated to the point of being repeated even by organizations such as the UN Environment Programme (UNEP 2002). It was actively revived around and after 1990 by analysts (Bailey 1989; Lomborg and Rubin 2009; Plenty of Gloom 1997) who subsequently dismissed *LtG* because depletion and collapse had not taken place. Other criticism included the claims that all scenarios ended in collapse, or that the models only run to 2040. However, the *LtG* authors never made these claims in the book, a fact that is easily verified. One would not even have to read the *LtG* books to check any of the above claims for accuracy. By simply looking at the books' graphs you can conclude that all scenarios run to 2100, they do not all end in collapse, and the ones that do, show that steep decline setting in after 1990. Yet, by the turn of the last century, these false claims had been so effective as to convince most everyone that *LtG* had been relegated to, as Lomborg and Rubin put it in 2009, "the dustbin of history". In a 2018 Harper interview (Ketcham 2018), one of the *LtG* authors, Dennis Meadows, describes how he met readers in the 1970s and 1980s who said the book had changed their lives. "In the 1990s and 2000s, they said, 'Your book changed my parents' lives.' Now," he said, "I give a speech and people ask, 'Did you write a book?'"

5.1.2. Technical Modeling Criticism

Some of the technical modeling criticisms were more on point, although they were not strong enough to classify as a refutation of the key *LtG* messages. Some technical criticism focused on the workings of World3 specifically, while others critiqued SD modeling itself as non-rigorous or even unscientific.

Then, there was the criticism of World3's sensitivity. Some researchers pointed out that relatively small parameter changes will in some cases significantly alter a scenario's trajectory (Castro 2012; de Jongh 1978; Vermeulen and Jongh 1976). It is true that World3 exhibits this behavior; however, that does not necessarily discredit the general validity of World3's outcomes. Sensitivity is problematic in predictive models because it reduces the confidence one can have in a prediction. But World3 is an SD model and, thus, not meant to be predictive. It is a tool to understand world dynamics, and as mentioned in Chapter 2, this world is full of tipping points, non-linear jumps, and other behaviors that make an altered trajectory in response to a seemingly small parameter change not that surprising. What would be important in an SD model is if the general dynamics remain accurate through parameter changes, not whether the timing of such events can be robustly predicted within a few years' precision (Lyneis 2000; Sterman 1994). And indeed, the recreation of runs with the same parameter changes as in these critical studies confirmed that World3 can be sensitive to parameter changes but also showed that these changes did not change the general dynamics of an overshoot and collapse pattern (Turner 2013).

Other technical modeling criticisms were less convincing. They came from acclaimed academics in their respective fields, but those fields did not include SD modeling. And that showed, as the criticisms seemed to be based on a lack of understanding of systems thinking. For example, a 1973 technical review of World3 concluded that it was inadequate from the perspective of linear modeling (Cole et al. 1973). This may be true, but that perspective is not the right criterion for SD models (Barlas 1996; Sterman 2000). It is a clash of mental models, not a criticism of any individual SD model; from a linear modeling perspective, nothing but linear modeling will prove adequate. Nobel-Prize-winning economist Nordhaus (1973, 1992) focused on the isolated equations of World3 in a response to the first and second *LtG* books, thereby neglecting feedback between system variables in his analysis (Forrester et al. 1974; Turner 2012). But of course, the feedback among variables is an essential part of SD modeling. As discussed in Chapter 2, the key and novel part of systems thinking and its modeling tools is precisely that interaction between a system's parts. Next to that criticism, Nordhaus (1973) also made a separate claim that "not a single relationship or variable is drawn from actual data or empirical studies" (p. 1157). This is simply incorrect. It is true that historical data are not fitted to a model using econometrics, but historical data were used to set the parameter values for the assumptions underlying World3.

There is an important difference between SD models and the econometric models that academics like Nordhaus and Cole et al. are used to, which I already briefly alluded to when introducing the accuracy measures. Econometric models assume some kind of constancy (e.g., linearity, homoscedasticity, independence, normally distributed errors, etc.), while SD modeling does not. Neoclassical economic models assume market equilibrium, whereas SD modeling does not. There is no recent evidence of why an (often unspoken) assumption of constancy and equilibrium is necessarily more scientifically valid than one that does not. Each of these types of models, with its different core assumptions, has its usefulness in the right setting and with intelligent application. But I would argue that in today's interconnected and thus dynamic world, models with strong implicit assumptions of constancy are not that useful for analyzing the dynamics of systems like a national economy or global society (e.g., Sterman 1994). Events like the financial crisis of 2008 (or the many others during the two centuries prior), and more recently the COVID-19 pandemic and social uprisings, painfully clearly demonstrate that sudden drops and jumps are part of the societal system's behavior. For any technique to be called adequate for modeling in those circumstances, at a minimum, would seem that it be able to account for such behaviors. Speaking of economics, let's continue with the *LtG* criticism using arguments from that field.

5.1.3. Economic Assumptions Criticism

Economic criticism of *LtG* mostly focused on a perceived lack of appreciation in the World3 model for technological innovation and price market correction. The first part, about humans' innovative capabilities, I have personally always found a bold one, to the point where I was tempted to put it in the section on misconceptions. I am of course paraphrasing here, but criticism by Cole et al. (1973) or Kaysen (1972), to me, seemed to come down to a much more eloquently put "but technological solutions, did you think about that?" That is quite a statement to make to a group of MIT researchers. More importantly, the *LtG* books did account for humans' ability to find technological solutions. As I already mentioned, some scenarios contained very optimistic assumptions about technological innovation and adoption, given historical averages. It is just that the *LtG* authors concluded that unless they were paired with societal values and policy changes, even the very optimistic assumptions on humankind's ingenuity and willingness to share solutions (also with those who cannot pay for them) did not prevent declines. The other common economic criticism, the absence of a corrective price mechanism, was more on point in the sense that it is true that money, and thus prices, are not part of the model. Critics contended that increased prices would spur substitutions among resources and other technological solutions (Kaysen 1972; Solow 1973). Nobel-prize-winning economist Solow (1973), for example, argued that increased scarcity would drive up prices of non-renewable resources, and pollution externalities would drive more regulation and higher taxes. There are two major counterpoints to this argument. Firstly, the absence of a market does not mean that there is no feedback mechanism at all. Again, in an SD model, variables directly interact with one another. Secondly, it is also not a given that market mechanisms will spur the necessary innovation and substitution rates to a sufficient extent. Indeed, we can now in hindsight conclude that they have not. Research by the IMF (Parry et al. 2021) and the OECD (2017, 2022a), among others, suggests that the social costs of pollution and non-renewable resource depletion are currently nowhere fully reflected in taxes. In fact, fossil fuels alone still carry large government subsidies, totaling USD 5.9 trillion, or about 6.8% of the global GDP in 2020.

5.2. Our Flawed Economic Mental Model

One reason that the economic criticisms turned out to be off is probably that they were mostly based on neoclassical arguments. Neoclassical economics has been the dominant school of economics for over a century, but it does not always describe well how our economy actually functions because one of its core tenets, the Homo economicus, does not describe human beings very well. The Homo economicus, or "rational agent", is self-interested and has fixed preferences and perfect information. This may be fairly accurate in some situations. For example, when choosing a toothbrush, we have all the necessary information: hardness of brush, color, price, and material. More than that, we don't need, and most likely, a person's preferences regarding these aspects are fairly constant. However, neoclassical economics assumes that we can each be represented at all times by this ever-and-only-calculating person. Since the first *LtG* publication, the Homo economicus assumption has been challenged, most notably by Kahneman and Tversky's (1979) paper on human decision making. This paper and their subsequent work on prospect theory detailed, quite convincingly, how human beings can behave very differently in practice from the theoretical rational agent, and won Kahneman a Nobel prize in Economics in 2002 (Tversky died before he could receive the same honor). A school that has been quickly rising in popularity since then is behavioral economics. This was very first started by Paul Samuelson in the 1950s but gained its real impetus with Kahneman and Tversky's paper, and was further developed by Sunstein, Thaler, and many more behavioral economists today. But criticism of neoclassical economics is not new, and you don't have to be a behavioral economist to voice it. Institutional economist Thorsten Veblen—who coined the term "conspicuous consumption", the phenomenon responsible for lines around the street corner of a shop selling "drops" from the hottest brands —is a late 19th-century example. A more recent one is post-Keynesian economist Steve Keen (2011, 2021), one of the few who predicted the global financial crisis (GFC). Economics students, young people that made a deliberate decision to learn more about that very field, have shown frustration over the gap between theory and reality. I am one of them, albeit retroactively. A better example is Harvard economics students walking out in 2011 on the class of Dr. Mankiw—Economics Professor and former advisor to President George W. Bush—to protest what they called teachings disconnected from societal reality, showing an "inherent bias" towards the rich (NPR Staff 2011). And although he still seems a supporter of the growth imperative (abandoning that is probably too tall an order for someone who won a Nobel prize with a growth model carrying his name), over the years, even Solow (2003, 2008) has become much more critical of the neoclassical school of economics.

Although further research is necessary (and indeed ongoing), especially among people who are not Western, educated, industrialized, rich, and democratic (WEIRD), recent research indicates that the notion of one standard rational agent just does not help us much in understanding how we can manage and improve our society. We are not much like that perfectly informed, calculating, and self-driven person. For starters, we don't always have perfect information. But even if we do, we don't always accept that information, especially not if we experience it as negative, or if it clashes with our current notions of the world. This is a common behavioral trait called motivated reasoning (Weir 2017) or the ostrich effect (Karlsson et al. 2009). And then, even if we do accept the information and know what the logical course of action is, we still don't always take that action. For example, Professors Abhijit Banerjee and Esther Duflo (Banerjee and Duflo 2019), who won the 2019 Nobel Prize in Economics, asked people in one research project whether an unemployed person should move to a different area for an available job, to which 62% of the respondents answered yes. But then, when people were asked whether they themselves would move if jobs in their community would disappear, only 52% responded affirmatively.

This figure dropped further when the respondents were in fact unemployed; as soon as the question was no longer purely theoretical, only 32% of respondents said they would move. The maximizing Homo economicus would always move for a job. Real people, however, are often reluctant to do so even if they realize there is little economic promise in their area, and they know other places offer more opportunities. They may feel rooted in their community. Or, they might be afraid of the unknown, especially if they have low self-esteem (from being unemployed in a society that tells us our worth is commensurate to our income, for example). The fact that these people are unhappy with their situation does not automatically mean they will make sensible decisions and take calculated risks to improve their long-term prospects. Neoclassical economics offers no explanation for the growing number of QAnon followers living in their mom's basement.

The assumption of fixed preferences also seems untenable against the background of recent social studies. First, preferences differ among people, based on an array of things, including genetics and upbringing (e.g., Kenrick et al. 2009). Second, preferences turn out to vary widely for the same person based on situational factors, including external pressures, power position, non-coercive hints or simply observing the behavior of people in their social circle (e.g., Kahneman 2011; Urbina and Ruiz-Villaverde 2019). Third, our preferences vary randomly, even in the same person in the exact same situation (Kahneman et al. 2021). None of these findings are compatible with the Homo economicus. Now, the first point we could still adjust for by letting go of the usual assumption in neoclassical economics of one single representative agent. By inventing a few rational agents instead of the single one, we might still be able to hold on to some neoclassical theories. But the invention of more than one rational agent might only adequately account for genetic aspects and much less for upbringing. What someone experiences throughout their lifetime is not just dependent on personal experiences within their family. It is also highly dependent on their class, race, gender, and sexual orientation. Neoclassical economics is ill-equipped to account for these effects because it almost completely ignores institutional factors. This brings me to the second point, namely that of people making different choices given the same economic risks and payoffs but under different conditions. Studies convincingly reveal that you and I will behave differently depending on whether we are under some form of stress. Financial stress, for example, significantly lowers a person's IQ (Mani et al. 2013). If the stress is temporary, a person's IQ can bounce back after the stressor is removed. But if there is permanent stress from living in poverty, this situation will reduce someone's ability to make long-term choices, i.e., precisely the kind of choices that could help the person climb out of poverty. Differences in brain activity between children who grow up in poverty and those who do not can be detected as early as infanthood and have recently been shown to be causal, as these differences reversed when the low-income mothers were given predictable and unconditional money transfers (Troller-Renfree et al. 2022). Experiencing violence or otherwise feeling unsafe can also change our ability to act like the independent, calculating, maximizing Homo economicus, especially when the stressful event(s) occurred while our brains were still developing. That is why, according to the Centers for Disease Control and Prevention (CDC 2021a), childhood trauma can lead to lasting "negative effects on (…) life opportunities such as education and job potential". Other examples of systemic stressors are racism, sexism, and heteronormativity, all of which have been shown to impair the health and cognitive functioning of those who experience them regularly (Coogan et al. 2020; CDC 2021b; Homan 2019; McDermott et al. 2021). But there are also many more benign one-off events can alter someone's stated preferences. Simply observing others, especially those with whom we identify, can strongly influence what we prefer at that moment (Thaler and Sunstein 2009). This phenomenon flies in the face of the notion of fixed preferences, even though it is obvious to anyone who works in marketing or has heard of social media influencers. Lastly, the third point, people behaving differently in the exact same situation is simply irreconcilable with the notion of the rational agent. A Homo economicus is constantly economically optimizing based on the latest available information. If the information does not change, neither would the rational agent's wants.

Finally, we make decisions based on a much wider range of social and economic factors than only self-serving ones. Research centered around how people behave, rather than the prescriptive theorizing of how they should, indicates that we are more altruistic and much more concerned with reciprocity than a Homo economicus could ever be (e.g., Haidt 2013; Kahneman 2011). We help or share with people at a personal cost, even under the certainty that a failure to do so would have zero consequences for us. We also go to great lengths to punish people we think have treated us unfairly, even if no financial or reputational reward can be expected from these actions. Plenty of alternative mental models have been suggested, such as Homo reciprocans (Bowles and Gintis 2002), Homo heuristicus (Gigerenzer and Brighton 2009), or simply Homo sapiens (Thaler 2000). My personal favorite is Homo duplex (Kluver et al. 2014), which proposes humans switch between two modes: one "lower" level where that person is self-focused on their goals and methods of achieving them, and a "higher" level where the individual mostly pursues goals of their collective group in a cooperative way. As I will explain in the next chapter, I believe that we will work mostly in the first mode when our physiological needs are not met but have a natural tendency to switch to the higher level mode once we are no longer resource-constrained. What all the research underlying these different proposed mental models has in common, though, is the conclusion that we are socializing animals rather than maximizing calculators.

None of this is to say we are irrational. If anything is irrational, it is to insist that humans should behave like this fictional Homo economicus in the face of a mountain of evidence that we do not. Therein lies the problem, because policies have been influenced by neoclassical economics. And that hasn't always worked out well. Take financial policies. Neoclassical economics boasts some impressive-looking mathematical formulas, but they heavily rely on the notion of a market equilibrium, which, in turn, often can only be derived if we assume one fully informed representative agent with fixed preferences. But because human beings are not even close to that paragon of all-knowing equanimity, markets are not nearly as stable as neoclassical economists and the policies they influenced assume. You are likely old enough to remember how the 2008 GFC shattered the idea that we had finally found the formula for financial stability, but this collective disillusion was hardly an unprecedented occurrence. Chief economist at the WB Carmen Reinhart and Harvard economist Kenneth Rogoff (Reinhart and Rogoff 2009) documented in their book *This Time is Different* that financial crises have been a regular phenomenon over the past 800 years. After a time of financial stability, people become confident that this period will last. "This time we have figured it out, this time is different".

It practically never is, argue Reinhart and Rogoff (2009). Inevitably, the financial market has a "Minsky moment". This term stems from the work of economist Hyman Minsky (1992). In his financial instability hypothesis, Minsky describes how investors will increasingly take on more debt during the stable period, introducing fragility into the financial system. This fragility increases with more debt and increasingly more speculative investments. When some of the risks materialize, there is a moment when people's overconfidence wanes, which is enough to set off a collapse in asset values in the now highly debt-leveraged financial market. Minsky's work only gained widespread attention after the GFC, a moment of recognition that Minsky himself was not able to witness because he had died twelve years earlier. We know that financial crises impact people's lives through the loss of a house or job. But the notion of a rational man has not just influenced financial policy; it also provided economic arguments that directly influenced socioeconomic policy. For example, the reasoning goes that if more people want jobs, employers (or at least the rational ones) can, and will, lower wages, and therefore immigration will depress wages. Similarly reasoned, raising the minimum wage will depress hiring. However, the empirical research of 2021 Nobel Prize winner David Card showed that immigrants don't necessarily lower the pay for native-born workers, and an increase in the minimum wage does not necessarily hinder hiring. This of course still does not tell us how to deal with immigration or whether to raise the minimum wage. But when dealing with important societal issues such as these and many others, let's at least base them on mental models more adequate than a fictional Homo economicus that does not represent us, invented for mathematical formulas that do not help us design more effective policies.

What it comes down to is that human beings are much more than just consumers and producers. We fulfill a plethora of social roles every day, and in most of them, we are much kinder than the neoclassical mental models would have us believe. That said, although our social nature might make us vulnerable to manipulation through influence, it doesn't make us weak; it is our greatest strength. As Brian Hare and Vanessa Woods (Hare and Woods 2021) detail in their book *Survival of the Friendliest*, Homo sapiens were able to thrive as we have not despite but because of this kindness. What gave us the evolutionary edge was not our analytical abilities but our empathic ones. The key message of their book is that we need to expand our definition of who belongs on this planet. Our survival might just depend on being this kind. As, on a smaller scale, it always has. But if we expand that definition of who belongs on this planet to say, people living in poverty, animals, and all other forms of nature, that will require society to share more equally. This brings us to distribution.

5.3. Distribution

I have given many guest lectures on my research at colleges by now, and my favorite part is always the discussion with the students afterward. Although I like receiving positive comments, the most valuable feedback has been in the form of critical questions. The most valid criticism in my opinion is that a global model lacks a distribution factor, while there are obviously large differences in how resources are allocated between people. This criticism is absent in the above overview because I have not found much record of it. I must assume it was mentioned by some when the first book came out, but it was clearly not the main criticism at the time. Perhaps it is understandable that income and wealth inequality (IWI) is more on people's minds today, as the difference between the rich and the poor within countries has grown since the 1970s (Chancel et al. 2021). So, it is true that IWI is an important force that should be part of any analysis of the (un)sustainability of business as usual. But if anything, the lack of distribution made World3 biased towards optimism rather than, as critics accused it of, towards doom and gloom. That is because inequality breeds inequality. Today's economic inequity constitutes a plethora of *Success to the Successful* dynamics, the archetype we discussed in Chapter 2. I will summarize here the various economic, behavioral, social, and environmental factors of IWI from a systems perspective. I will focus on the US because, after having lived and studied here for eight years, I am most familiar with it by now, and IWI in the US is the starkest compared with other developed economies. Nevertheless, although the level of IWI and specific details of its impact can differ for each country, the general dynamics can be observed in most other economies too.

5.3.1. Economic Factors

IWI in the US has been steadily growing for the past 50 years (Schaeffer 2020). The middle class has not seen a real income increase in decades. Most of the modest real growth in US median wages that can be discerned is accounted for by increases in pay for both women and people of color as they have slowly been catching up with white male workers (USCB 2021a). A 2021 poll found that the majority of Americans, three out of five, wake up in the middle of the night sometimes over financial worries (Melore 2022). Even before the COVID-19 pandemic, 40% of Americans would struggle to afford a USD 400 emergency expense (Board of Governors of the Federal Reserve System 2019), a quarter of the population skipped necessary medical care because they could not afford the cost, and one-sixth were unable to pay all their bills in full every month. According to the US Census Bureau, more than 37 million Americans (over 11% of the population) were living below the poverty line in 2020 (USCB 2021b). According to the US Department of Housing and Urban Development (2021), more than half a million Americans were homeless in 2020. And according to the United States Department of Agriculture (2020), about 6.5% of households with children are food insecure to the extent it has negatively impacted the children's diet. In the richest country in the history of the world, that is a lot of people living under immense financial stress, without shelter or sufficiently varied and nutritious food for even their children.

Part of the reason for this stagnation in living standards for the lower and middle class in a country that has been generating ample income is the way that global finance today works (Piketty 2014). The lower- and middle-class workers receive most of their income from labor, while the wealthy do so mostly from investments. Now consider that, in a globalizing world, investments provide higher returns than labor. Financial flows are flexible in terms of location; it is relatively easy to reallocate money from one continent to another. Labor, on the other hand, is most often solely delivered domestically, especially for blue-collar workers. Financial returns are higher in emerging economies than in developed ones, so countries with promising economies attract foreign investments in this situation. This then comes at the expense of domestic investments in the matured economy. Globalization thus benefits the wealthy through higher returns, and, under certain conditions, which include the financial flows not being too fickle, also the workers in the emerging economy. But in more developed countries, like the US, the drain of investment funds from their economy has led to a stagnation in real wages. Investment providing higher returns than labor is another example of the *Success to the Successful* dynamic, for financial capital, as depicted in Figure 25.

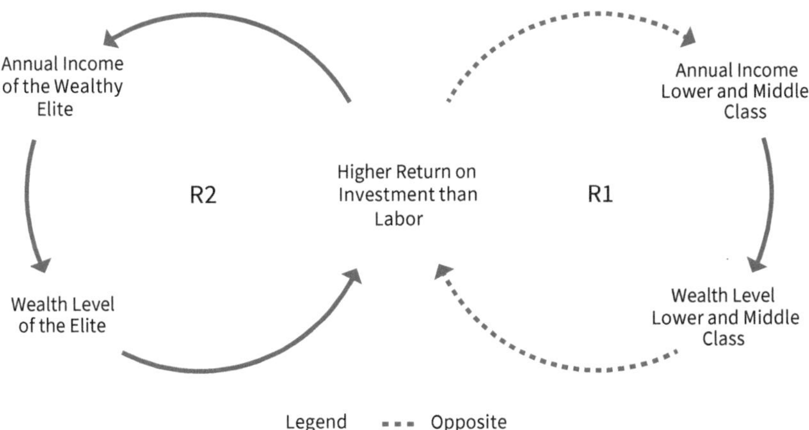

Figure 25. The Success to the Successful archetype for financial capital. Source: Figure by author.

A 2020 study from the RAND Corporation found that since the mid-1970s, around USD 2.5 trillion per year has been redistributed from the bottom 90% of income earners to the top 1% (Price and Edwards). If wages had kept pace with inflation and GDP, a median prime-age full-time worker earning USD 50,000 annually in 2020 would instead have had about double that salary.

Globalization is not the only factor that has contributed to this wealth transfer, although it is a key one. Other economic factors include technological change and the decline in unions. But these form their own self-reinforcing dynamics. As Stanford and MIT professors Erik Brynjolfsson and Andrew McAfee (Brynjolfsson and McAfee 2014) put it in their book *The Second Machine Age*, "the technology-driven economy greatly favours a small group of successful individuals by amplifying their talent and luck". Simply put, if you own the robots, you'll reap the rewards of their increased productivity; if you don't, you'll be replaced by them. Unions are less straightforward. You would think that eroding real wages would automatically spur increased efforts from lower-class workers to unionize and demand more. But that's not necessarily how people work. The e-commerce giant and second-largest US employer Amazon, for example, has indeed faced increased unionization efforts from warehouse workers over the past few years (Rubio-Licht et al. 2022). But until only recently, these efforts had failed. A widely covered attempt at unionizing in Bessemer, Alabama, for example, was voted down by those very low-wage workers. In April 2022, an Amazon warehouse in Staten Island, New York, historically voted in favor of a union. Since 2021, over 220 Starbucks locations have voted for unionization at the time of writing, which is about 1% of all US stores. However, according to labor and workplace researchers (e.g., Greenhouse 2022), it is far from certain that these are signs of organized labor making a comeback, rather than a few small victory exceptions. That is because another dynamic is at play here aside from just the economic one. This brings me to IWI's behavioral dynamics.

5.3.2. Behavioral Factors

Contrary to the ruthlessly calculating rational agent at the heart of neoclassical economics, social research on human beings is quite clear on the fact that we are hard-wired for fairness (Haidt 2013). When this need for fairness is not satisfied, social unrest can follow, or put more simply: people get angry. This anger will not always be directed toward the root cause of people's disenfranchisement. Crime rates show positive correlations with inequality (Rufrancos et al. 2013) much more than they show any correlation with the severity of punishments (Fajnzylber et al. 2002; National Research Council 2014). Less extreme manifestations of social unrest come in the form of tensions around ethnicity, gender, religion, or sexual orientation. Indeed, a Princeton study found a link between inequality levels in US states and individuals' propensity for "racism, sexism, welfare opposition and even willingness to enforce group hegemony violently by participating in ethnic persecution of subordinate out-groups" (Kunst et al. 2017, p. 1). Then there are the mundane, everyday interactions. People living in economically unequal US states demonstrate overall lower levels of agreeableness and show more propensity to display a competitive, less trusting, more self-focused mindset, as opposed to a default focus on cooperation and reciprocity (de Vries et al. 2011). In short, high inequality seems to stimulate the kind of behavioral attitudes that not only increase people's tolerance for inequality but sometimes even lead them to actively maintain it.

The lack of trust is not exclusive to between-citizen interactions. According to the IMF, generalized trust within the US has been steadily declining for decades, both in others as well as in the government (Gould and Hijzen 2016). About 44% of this decline was estimated to stem from increased inequality. In fact, the US public's trust in the government is near historic lows (Rainie et al. 2019). Government has the legislative power to counteract IWI directly, for example, with tax and corporate regulations, as well as indirectly with universal health care and public school funding. However, higher IWI also makes it easier for the economic elite to influence politicians into defending their interests (Krieger and Meierrieks 2016; Piketty 2014). An atmosphere of reduced trust amongst civilians creates room for a rise in populism (O'Connor 2017). IWI thus erodes democracy simultaneously from both top down (influencing of politicians by the upper class) and bottom up (resentment of the political establishment in the lower class). A recent prime example of this is the storming of the US Capitol on 6 January 2021.

5.3.3. Social Factors

An abundance of academic research suggests that the lack of trust and social cohesion that results from inequality has a detrimental psychosocial impact on health (Ellison 1999; Rözer and Volker 2016; Bergh et al. 2016). However, IWI affects people's health further. Low- and middle-income Americans have a lower life expectancy than their upper-class peers (McGinnis 2016). In general, life expectancy for Americans is at its lowest point in decades (McPhillips 2022). The sharpest declines seen recently can be explained by the COVID-19 pandemic, but life expectancy had already leveled off for a decade before the pandemic began (WB 2021i). Life expectancy is largely considered a proxy for a population's health, although it is not the only one of course. Still, the stagnating life expectancy could suggest a larger trend of economic growth not improving Americans' health status anymore, at least not for the majority. Indeed, there are significant inequalities in health status in general, and the myriad of influences contributing to these inequalities can be aggregated under two major predictive indicators: income and education (McGinnis 2016). In other words, the higher someone's income and education level, the more likely it is that their health status is also relatively high. The influential factors that contribute to the overall health disparity related to income form a wide range, only some of which are the unavailability and/or unaffordability of nutritional food products (Otero et al. 2015), more exposure to toxins, air and water pollution (Allaire et al. 2018; Dodson et al. 2017; Ruiz et al. 2018), the opioid crisis (Harper et al. 2021), and lower-paying jobs rarely offering adequate maternity and sick leave policies (Jones 2017). These influences are exacerbated by the fact that households with lower incomes are less likely to have health insurance and access to quality health care in the first place (USCB 2021c). But along with wealth and income levels, the other predictive indicator, education, also has also become less equitable (Eide et al. 2010; Duncan and Murnane 2014). In *Degrees of Inequality: How the Politics of Higher Education sabotaged the American Dream*, Suzanne Mettler (2014) describes how a combination of steeply increased tuition fees, political capitulation to corporate interests, and reduced disposable income for most of today's parents has many students leaving colleges with massive student loan debts and little more opportunity to show for it. The above interactions were mentioned in Chapter 2, as indicated in the first example of a *Success to the Successful* dynamic in Figure 5.

5.3.4. Environmental Factors

A more equal income distribution in the US has been found to have a beneficial effect on the environment (Baek and Gweisah 2013), while IWI in a country seems to worsen environmental sustainability (Islam 2015). Here, too, environmental neglect works from both bottom up and top down for behavioral reasons. People who are struggling behave in reactive manners; they focus on immediate concerns, whereas the longer-term ones, like protecting natural capital, take a backseat (Dorling 2017). Research also shows that upper-class citizens of the more unequal developed countries on average consume and waste more than upper-class citizens in more equal developed nations, for reasons which include peer pressure to display a certain lifestyle (Dorling 2017). During the 2012–2014 drought in California, for example, some wealthy people hired trucks with water to come in to maintain their green lawns, while everyone else let their lawns go brown to conserve water (Bardach 2014; Pincetl and Hogue 2015). As economist Magnani (2000) put it, "income inequality produces a gap between the country's ability to pay for environmental protection and a country's willingness to pay" (p. 431). On top of this reduction in overall willingness to protect the environment, a community has a harder time coming together to manage the commons responsibly because inequality damages the social fabric (e.g., Nair 2018).

In the long run, this ecological apathy will cause a loss of biodiversity, deplete natural capital, and contaminate the environment. As mentioned above concerning health, worsening environmental conditions often disproportionally affect the poor, consequently raising inequality further. This works both in terms of availability and affordability because the increased scarcity of uncontaminated natural resources may cause their prices to go up (WB 2012).

5.3.5. Rich to Riches

Many of the earlier described dynamics constitute a *Success to the Successful* archetype; I don't need to draw a CDL of each one. And of course, IWI is a much more complex problem than just a collection of archetypes. Figure 26 below, for example, is an attempt to connect relevant factors more comprehensively.

The above-described elements and their interplay culminate in family wealth lineages superseding any American's hard work, talent, and innate intellectual abilities when it comes to the expected level of future wealth (Pfeffer and Killewald 2018). This flies in the face of the promise of meritocracy. Social mobility is lower in the US than in most other high-income countries in the world (Dabla-Norris et al. 2015). This means that the American Dream, working hard to move ahead and climb the societal ladder, today comes true more often in other parts of the world than in America. In 2017, a UN special rapporteur, who visited the US for his research on extreme poverty and human rights in the country, concluded in a statement shortly after his visit that the "American Dream is rapidly becoming the American Illusion" (Alston 2017, para. 12).

On a global scale, inequalities between countries show the same kind of dynamics as described above. They can become more complex, for example, because of foreign investments. It should also be noted that, unlike income inequality within countries, global inequalities between countries have declined over the past two decades (Chancel et al. 2021). Still, the same general dynamics among environmental, social, and economic factors are observable. For example, richer nations contribute much more to climate change in terms of per capita carbon emissions, but the effects of climate change disproportionally fall on other nations (Chancel et al. 2021; Eckstein et al. 2021). The economic inequalities are accompanied by the familiar racist and sexist inequalities. The poorest countries in the world are all in Africa, except for Afghanistan. Women own half as much of the world's wealth as men (Zakrzewski et al. 2020). According to Oxfam International (2020), the 22 richest men in the world have more wealth than all the women in the entire continent of Africa combined. The COVID-19 pandemic has only exacerbated the existing economic and social inequalities (Ahmed et al. 2022). In short, on a global as well as individual scale, although our individual choices do matter, they are often outweighed by the relevance of where we are born (Roemer 2000).

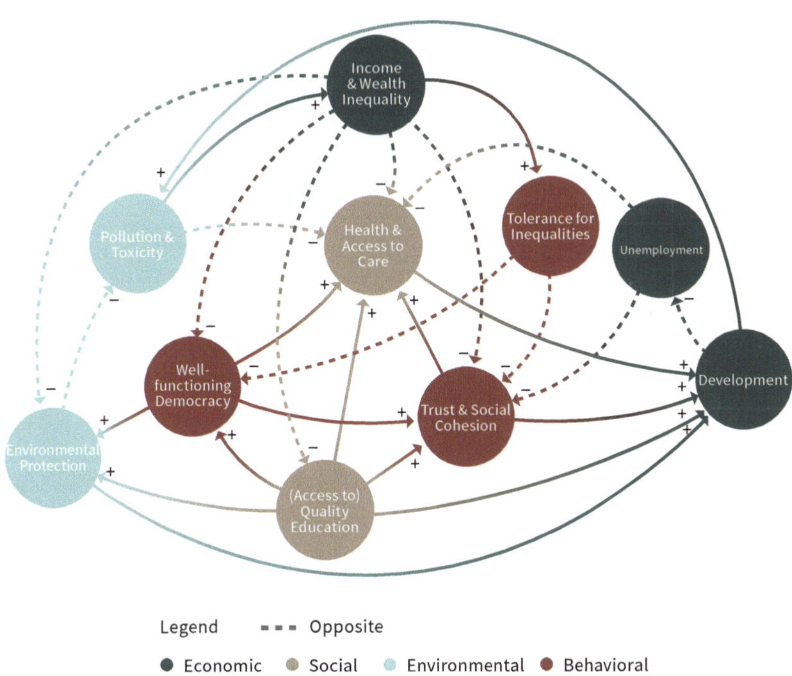

Figure 26. CLD of behavioral, social, economic, and environmental factors of US IWI. Source: Figure by author.

5.3.6. A Rising Tide Can Drown Us All

The above findings may seem unfair; however, you might have noticed that the arguments in defense of inequality are also based on a notion of fairness. Combatting IWI would have to entail policies like high taxes on wealthy individuals and corporations, among other things. But, the protest goes, people worked hard for those incomes or profits and accumulated wealth, so it is unfair to take it away from them. At best, people making these arguments seem to have a poor understanding of system dynamics. The implicit assumption underlying this type of fairness argument is that everyone has an equal chance. But the findings cited earlier show that we do not. As Yale economist John Roemer (2000) points out, fairness means equality of opportunity. This cannot be achieved without a level playing field that is actively maintained. Even if a society starts out as a pure meritocracy, it will not stay a meritocracy for long unless the countless *Success to the Successful* dynamics are kept in check.

Sometimes, people who argue for redistribution are accused of being sore losers and simply envious of the rich. I find this a weak argument, first and foremost because it is a personal attack and does not address content. But if we are getting personal, then let me add my little note on this: I do not want a yacht. I could die very happy without ever having found myself caught up in the conspicuous consumption of a USD 33,000 Gucci White Tee (in case you were wondering: no, that is not the most expensive t-shirt in the world—that one costs USD 400,000). Reasonable people can disagree about what is an optimal tax rate, or in general, what constitutes a level playing field. But one's argument should be based on more than the assumption that the other person is jealous of some billionaire's gold-plated toilet.

The arguments most often heard in defense of IWI are neoliberal. Neoliberalism and neoclassical economics are not the same: The first is a political ideology, whereas the second is an economic framework. They do seem to make good bed fellows in practice; a person arguing for neoliberal policies will often use neoclassical economic arguments. But the rational agent at least can come out in specific settings, like the toothbrush example I mentioned earlier. A market can, under specific circumstances and for a period of time, be in equilibrium. Despite often being brought with a confidence I am in fact truly envious of, key neoliberal theories about the benefits of small government and low taxes have little meaningful empirical evidence to support them. Moreover, much evidence points to the contrary. Small government, for example, according to neoliberal theory, would allow for more innovation because the corporate sector is supposedly more entrepreneurial. Therefore, leaving it unhindered by government regulation and intervention will boost economic development. However, the biggest economic developments over the past few decades have not occurred in countries with the smallest governments. The region with the highest growth over the past few decades has been Asia, especially China. These "Asian Tigers" are hardly paragons of neoliberalism; they have each had their own version of a state-led economy, especially, again, in China (e.g., Movahed 2019; Nair 2018). Additionally, the state has been quite the catalyst for many innovations in history, also in the West. As best-selling economist Marianna Mazzucato (2015) details in her book *The Entrepreneurial State*, many of what we today consider the most innovative products were made possible through government-funded research. Everything that makes your smartphone smart, including the internet, GPS, touchscreens, and voice recognition, exists thanks to government-funded research. It may have been business owners who shot a rocket into space in 2021, but the National Aeronautics and Space Administration, the US government agency better known as NASA, was already launching rockets decades earlier, in the 1960s. Another neoliberal theory is that an open market automatically benefits both trading partners. But it has hardly always worked out that way for developing countries that opened up their young economies to international trade (e.g., Chang 2011). Indeed, most of today's developed countries that have been arguing for open market policies in developing countries only opened their own economies after they had matured during a period of strong protectionist policies.

The prime neoliberal argument for high IWI is of course trickledown economics. This argument cautions us against any redistributive intervention because wealth at the top will automatically find its way down to the rest of the income levels. Perhaps it won't erase inequalities, goes the reasoning, but you shouldn't want that anyway because the people at the top spur growth which translates to benefits for everyone. We should remove any obstacles for these "job creators" and let a rising tide float all boats, as it is put poetically. This theory is appealing, as everyone likes a win–win, but it lacks supporting empirical evidence. Ignoring for now that the trickledown economics argument equates growth to benefits for all, even in developed economies, inequality does not even deliver on the promise of growth itself in the long term. Tax cuts for the rich are uncorrelated to job growth (Stiglitz 2012). A 2015 IMF Discussion Note, based on multi-country econometric analyses, concludes that (Era Dabla-Norris et al., p. 4): "[as] the income share of the top 20 percent (the rich) increases (…) GDP growth actually declines over the medium term, suggesting that the benefits do not trickle down". Federico Cingano (2014, p. 6), a researcher at the OECD Economics Department, concluded after his own study that "income inequality has a negative and statistically significant impact on subsequent growth". In a paper called *Neoliberalism: Oversold?* (Ostry et al. 2016, p. 39), the IMF research department wrote that the benefits of neoliberal policies "in terms of increased growth seem fairly difficult to establish (…)". In contrast, high tax rates for high incomes do not seem to impact a country's economic performance (Stiglitz 2012). There is no ebbing tide that strands all boats discernible in the data. On the contrary, as Piketty's 2022 book *A Brief History of Equality* details, it was the advent of progressive taxes on income and wealth, which paid for education, health care, and old-age pensions for all, that was the main reason for the significant increases in living standards that we have seen in some parts of the world. In the US, for example, the top marginal rate of personal tax was at least 70% from 1936 to 1970 (Tax Policy Center 2020). This period is widely recognized as America's heydays. Since then, a massive transfer of wealth has been made from the lower and middle class to the ultra-rich. The total transfer over the past three decades is estimated at USD 47 trillion (Price and Edwards 2020). This is a number none of us can comprehend. Our brains are hopelessly ill-equipped to deal with that kind of scale. Perhaps, though, we can try to imagine how different the US would look if that wealth, and the power that comes with it, had stayed with the majority.

5.3.7. Inequality in a Global SD Model

In a nutshell, a bit of research into distribution makes it evident that high IWI is bad economics and even worse social organization. Admittedly then, this could be an important factor left out of a global SD model. In that light, it is interesting to note one World3 variable that does not align so closely with empirical data: food per capita. It is clear from my comparison that food production has risen more and for longer than in any of the World3 scenarios. But then why did one out of every ten people still face hunger in 2020 (FAO et al. 2021)? In fact, after steadily decreasing for decades, the number of undernourished people has started to rise again over the past few years. The answer can only lie in distribution. After all, we have enough to feed everyone on the planet, and despite rising hunger in some regions of the world, we do throw away about 17% (900 million tonnes) of consumer food products globally, and more than double that percentage in the US (UNEP 2021a).

So, it does seem that distribution could be a relevant explanatory variable. This is one of the reasons why Randers, one of the *LtG* authors, chose to differentiate between regions when he, together with other researchers, set out to build a new global SD model. He also included a variable for "social unrest", which would increase along with inequalities in the system. This model, called Earth4All, is further discussed in Chapter 7. Yet at the same time, research into distribution seems to point to a conclusion not so dissimilar from that of my *LtG* research: The narrow pursuit of growth, without regard for social and environmental impacts, is unsustainable. This would imply that if anything, the lack of distribution in World3 means we have even less, not more, time to act.

5.4. The Urgency of Our New Common Narrative

The message that the time to act is now in this context has of course been voiced for a few decades. What makes the urgency real this time? Let's put the *LtG* message in the right perspective: The urgency was always there, but the consequences have changed. When the *LtG* authors and many other scientists started to sound the alarm around resource scarcity and environmental pollution in the 1970s, their message was that acting now would allow society to transition to a more sustainable society without many financial, social, and environmental costs. But around the time of the

30-year update to *LtG*, the authors had become "far more pessimistic than (…) in 1972" and concluded that humanity had "squandered the opportunity to correct our current course" (Meadows et al. n.d.). The *LtG* books were not, as Kaysen put it in 1972, a model "that printed out W*O*L*F*". My research indicates that the authors' message was accurate. Because it was not heeded, the opportunity for a relatively smooth transition into some form of global equilibrium is gone. The ride will be bumpy, and the only question left is whether we will make it at all. The warning has changed from "if we don't act now, our children's children will have lower well-being" to "everyone under 40 today will live an unprecedented life in terms of their exposure to heat waves, droughts and floods" (Chow 2021). In the meantime, preventable human suffering, biodiversity loss, and ecosystem damage have occurred and cannot be undone. We have definitively left the Holocene—a time of unusually stable climate on Earth which allowed Homo sapiens to thrive as we have—and entered the Anthropocene: the age where the biggest impact on Earth comes from humans (Lewis and Maslin 2015). No one really knows what this new geological era will bring, but it is bound to be less stable than in our past. And it is not just our climate that has become more unpredictable. Our social environment has become more unstable too, as a result of, among other things, economic and social inequities. This environmental crisis on the one hand, and social crisis on the other, are today's twin key converging challenges. These two macro developments result from the structures of our society, themselves resulting from our vision for humanity. We will not reverse them by tweaking at the margins of our structures, certainly not in the little time we have left to act. To make the transformational changes we need, we must operate at the generative level.

5.4.1. Why Did We Not Act before?

Humans are thought to have been so successful compared with other animals because of our unmatched ability to collaborate (e.g., Haidt 2013; Harari 2015). No other animal works together in groups of millions, but we do, with the help of shared fictions. A country is a shared fiction. Another currently popular one is the multinational corporation. My marriage is a story. My husband and I put on our wedding rings every morning before heading out the door. It feels important to do, even though I doubt anyone with whom we interact during the day would notice if we didn't. Every year, we make the effort to celebrate that day in December when we played dress-up, me in tulle and a tiara, and pretended to be merging kingdoms. We know nothing is particularly special about that day; we are aware that we can go out to a restaurant any other night. We make the effort to go on that specific date anyway because we enjoy the reaffirmation it brings to our shared fiction—the story that adorns the interactions between my husband and I, from our shared daily routines to the sparse lovely holidays that we always say we should do more of but never do. In short, our marriage is the narrative underlying the system my husband and I form. I like this system, so I'm going to keep the story. But plenty of people end the narrative of their marriage with a divorce every day. It is good that they have this option, because why suffer for a story? Fictions are not real in the sense that they only exist in our minds, and when they stop existing there, the stories themselves do not suffer. But as long as we believe a narrative, it can have real consequences, which might include the suffering of living things. A multinational may be a shared fiction, but the annual carbon emissions attributed to the company are real, as is the impact of these emissions on global warming and the suffering it has already caused to humans and other animals. Because, as discussed in Chapter 2, if we behave as if the fiction is real, we construct systems structures on those narratives. That is why it's important to occasionally check if a story is still serving us.

Money is the most successful shared fiction (Goldstein 2020). We have all been raised in the economic framework, so it feels real, but the capitalist narrative underlying our economic system is a shared story too, which becomes clear as soon as we replace the systemic threat with a purely physical one as in the Netflix movie "Don't Look Up" or the cartoon in Figure 27 below.

Figure 27. The comet that causes the extinction of the dinosaurs. Source: Figure created by Hilary Moore.

Yes, people suffer when, for example, they lose their job during an economic downturn. But to what extent does this suffering result from the event of the job loss, and to what extent is it a result of the way we have designed the economic system? What if someone's health, shelter, and family's safety were not impacted by the job loss, for example, through the availability of universal health care and shared dividends from the commons? If not every part of one's survival and identity depended on a job, would the suffering still stay the same? These questions are being asked more and more. When the *LtG* team came out with their first book in the 1970s, their message was quite clear in that the biggest changes would have to come from the rich countries and wealthy individuals. I believe that this might be one reason their message was met with so much aggression; even gradual change has the potential to upset existing power structures. The story of capitalism, which has the concept of growth at its very core, experienced a temporary dip in support in the 1970s when post-World War II growth began to slow, and *LtG* became a bestseller (Foroohar 2016). But support increased again once economic growth rebounded, in no small part due to deregulation of the financial sector (Krippner 2012). Since then, governments have become poorer, and private wealth has become concentrated further. These conditions can work out either way; in some ways the majority has even less power today, but on the other hand there are also more people with little to lose from changing existing societal structures. We don't know if this time, the necessary systemic changes will happen. But it is clear that today, the capitalist story is losing support again.

5.4.2. The End of the Current Capitalist Story

"If wealth was the inevitable result of hard work and enterprise, every woman in Africa would be a millionaire" (Monbiot 2011).

Facing mounting environmental pressures on one side, and social tensions on the other, the capitalist narrative has been losing ground. Edelman (2020), which has conducted polls among thousands of people all over the world about their trust in core institutions for over two decades, concluded in a recent Trust Barometer publication that the "majority of respondents are losing faith in the capitalist system". According to a recent survey among G20 countries, three in four people are aware that Earth is approaching tipping points and support deep systemic changes towards prioritizing health, well-being, and the planet (Gaffney et al. 2021). Another survey by the Pew Research Center reveals that the majority of people in advanced economies want deep reforms in their political and economic systems (Wike et al. 2021). In the US, this was at least two out of three. A majority of US voters, including Republican ones, see income inequality as a problem and support raising taxes on the wealthiest Americans (Casselman and Tankersley 2019). And American youth today is more often positive about socialism than about capitalism (Newport 2018). In Japan, a "Marxist, post-capitalist, green manifesto" became a bestseller thanks mainly to the interest of younger people (McCurry 2022). An English version of this book, *Capital in the Anthropocene*, in which the author Kohei Saito advocates for things like shorter working hours, sharing wealth, and prioritizing caregiving, is expected to come out next year. For too long and for too many people, capitalism, in its current form at least, has not delivered. The narrative can no longer hold us together. The wealthy elite is aware of this too. Some of them are acting mature, putting their wealth towards improving things for the global society here on Earth. But more than a few doomsday preppers are buying "billionaire bunkers" in New Zealand (Carville 2018; Hollingsworth 2020). As much as a hippie cliche my first insight that we are all connected must seem to some, this basic principle evidently is still not understood by all. As Jared Diamond (2011) describes in the book *Collapse*, throughout history, the wealthy have tried to use their power to protect themselves in case of a societal collapse. All they ever bought was the privilege to perish last.

There is hope, though, as shared fictions work a lot faster than evolution. We can quickly change our behavior by adopting a new story. We are typically reluctant to do so; as mentioned, we have a cognitive resistance to accepting information that doesn't fit with our existing narrative. But we can, and today, the capitalist story seems set to mutate into something new (e.g., Mason 2017). That new story is emerging; one of the many platforms where this can be observed is the World Economic Forum's *New Narrative Lab*, for example (WEF 2022a). But our new narrative is not here yet. And we need one urgently. My research, and that of many others, makes clear that we need a sustainability revolution. That is not a set of changes within old paradigms; it is a societal upheaval. Climate change experts have been warning of tipping points in the climate system, but there are also social tipping points (Tàbara et al. 2018; Otto et al. 2020; Westley et al. 2011). A social tipping point is a group dynamic where, at a certain moment in time, group members rapidly and dramatically change their behavior by widely adopting previously rare paradigms and practices. Such a social tipping point can still bring about the transformations of societal priorities which, together with technological innovations specifically aimed at furthering these new priorities, can bring humanity back onto the path of the SW scenario. If we can come together to form the new story for humanity, we can still make that turnaround. But will we, and if so, what will this new story look like?

6. Prosperity over Growth: From "Never Enough" to "Enough for Each"

6.1. A New Narrative

At first, humanity was no different from other animals. We evolved with them and with the nature around us. For a while, we were sure of our role, in our tribe and in the web of life. Then a spark gave us the power of speech. We made stories and so became able to adapt more quickly than the rest of the world. Unaccustomed to our newfound power, we were now unsure of our role. "If we take control", we thought, "surely we will be free and happy". And so, we used our power to dominate that which was soft and kill what was strong. "Look what I can do!", we roared. But the only sound that came back was our screen-echoed roar. The rest of the world stayed silent. And we felt lonely.

We asked organized religion what to do, but they gave the same old answers. We asked scientists, but they only told us how, not what or why. We accessed our own internal truths, revealing answers that were ever-changing.

Still unsure, we looked further for help. Then, we saw the brothers and sisters we had left behind shortly after the spark. Their numbers had dwindled since then. They looked odd to us, whispering to animals, trees, and water. But we had no other options left, so we asked them: "What do we do now?"

They said: "What do you need?" We thought about it. Someone answered:" I need money." They said: "I hear you need security. I need that too." Someone else answered: "I guess I want money because I need to buy a house and car." They said: "I hear you need shelter, a place in the community, and freedom. I need that too." Then we answered: "I guess I just need to feel safe and connected." They said: "I need that too.' And we then recognized they were us.

People started reconnecting with their needs, and then they recognized those in others. Some did this faster than other people. The fast adapters used their spark to guide those who had difficulty recognizing the needs of people who did not look like them. Although these guides showed no tolerance for aggression, they made sure to listen to the hurt underlying the anger of the slow adapters. And over time, the anger dissipated. Once we had reconnected around our needs, we could satisfy them. Bewildered that we had forgotten our birthright of loving every part of who we are, we now realized we were so much more than we had been telling ourselves.

This healed some of our internal wounds. But the world was still damaged, and therefore so were we. We asked our newfound global tribe: "What do we do?" And we answered: "Now, we listen to the forms of life which have been given a different kind of spark." And we started asking: "What can I do for you?"

We listened to other animals. We recognized that their needs for food, shelter, connection, and respect were like those of humans. And in the process of freeing them from the places we had kept them locked up, sharing our land with them, and re-establishing their habitats, we learned the deep responsibility that comes with our power.

We listened to trees, other plants, fungi, even those life forms so small we couldn't see them with the naked eye. We stopped clearing them and gave them space. We helped them grow when necessary and watched in awe as their boundless capabilities to store the Sun's energy, provide nourishment, and clean up our waste products were revealed to those who paid attention. We then knew enough to sustain every human being on the planet. Grown-ups with an ever child-like wonder, now we were finally ready to understand nature's essence as generosity. And thus, because we recognized we were She, the essence of our responsibility too.

"What about the ground, water, or air?", said someone, "that's not alive". "No", said another, "but they are life-giving". We understood that air, ground, and water cannot fulfill their roles if they are polluted. So we redesigned the way that we fulfill our needs to not harm the planet and to give back more than we take. We watched Earth's many cycles revive. And we experienced the fulfillment that comes with our responsibility to protect, restore, and connect.

"What about the chemicals we produced?" asked one. "Or fossil resources?" said another. By now, the answers came to us quickly: "Those are not alive, so they only have a place flowing through society if they are truly life-giving." We used our technology, in collaboration with nature, to clear up the dead and harmful things or keep them stored away. We gratefully accepted the energy from our openhanded Sun. Now humanity had become a regenerative force.

We marveled at how nature unfolded upon us, embarrassed that we ever acted as if we could live without life. Why did we ever think we would want to? Humbled, we fell at the riverbank. We watched the water, so strong as to cut through stone, so soft as to move without hurting, and so powerful as to give life wherever it goes, all at the same time. We promised water to never undervalue her again. Then the flood came and washed away our last regrets.

Things are not perfect. Tragic accidents still happen, hearts are still broken, loved ones still pass away too soon. There are still plenty of misunderstandings and genuine disagreements about many things. But we are no longer alone, no longer unsatisfied, no longer unsure of our role as Gaia's guardians. We needed to adapt, and we did. Because of that, our future has forever changed. And it is as it should be.

6.2. "Technology Can Save Us"

Over the past few years, I have given many presentations and speeches at conferences. Almost invariably, there will be someone who comes up to me afterward claiming that I underestimate human ingenuity and that we can all relax because the upcoming years will reveal new data showing the closest alignment to CT instead of

BAU2. My experience is hardly unique. Sandrine Dixson-Declève, co-president of the Club of Rome (CoR), gave a TED talk in October 2021 about what the message of *The Limits to Growth* means today for how society needs to transform (Dixson-Declève 2021). (She mentioned my research briefly at the beginning and described me as "very young" and "brilliant", which is definitely too generous on both counts.) Later, during a CoR subcommittee meeting, Sandrine shared how she and fellow CoR member Per Espen Stoknes were approached afterward by people who told them that humanity doesn't really need to change its behavior nearly as drastically as she stated because technology was obviously going to do most of the necessary transformation for us. I have no doubt that every environmental advocate has heard this argument.

At this point, both the CT and BAU2 scenarios are the closest fits. The two scenarios cannot effectively be distinguished because they haven't diverged sufficiently yet. But because they do show very different developments about five years from now (Figures 12 and 28), many people's understandable reaction to this result is to wonder which scenario humankind is more likely to follow. BAU2 depicts a scenario where pollution will cause societal collapse, while CT shows only a moderate and to some extent temporary decline in welfare levels.

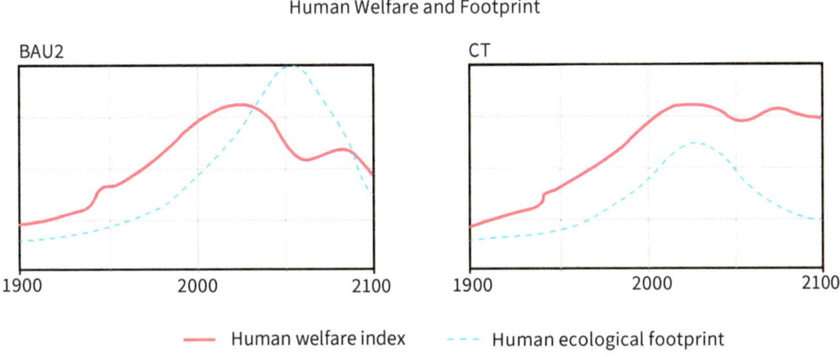

Figure 28. Welfare and EF developments for BAU2 (**left**) and CT (**right**). Source: Adapted by Hilary Moore from Meadows et al. (2004).

"Given the accelerating technological progress we're witnessing", I hear sometimes, "it's likely that we're following CT and so we should be fine with just a temporary dip around 2050, right?" The short answer is that, at the moment, we do not know. We could be following either scenario, or a mix of both, or neither. Currently available data are inconclusive, and no one knows the future with certainty. The longer answer goes as follows.

The question of what a best fit with both CT and BAU2 means comes down to whether we believe society could be heading towards collapse, or whether we will be able to mostly keep innovating ourselves out of this future. There is also the possibility of following neither scenario, but as I mentioned in Chapter 3, *LtG* updates have confirmed both the validity of the model's causal links and a close track with empirical data, so World3's output should not be dismissed that easily. The fact that both the BAU2 and CT are the best fits could also suggest a mix of the two scenarios: Those who can capitalize on climate mitigation and afford climate adaptation would experience more of a CT trajectory, while those who cannot do so would experience a situation more closely resembling BAU2. This would be in line with general inequality dynamics, as discussed in the previous chapter. The findings of Pasqualino et al. (2015) that humanity had invested more to abate pollution and increase food productivity, compared with BAU2, also support this possibility. On the other hand, that's not to say the study proves it. Remember that this study did not include a comparison to CT (nor SW), so we do not know if these efforts have been sufficient to bring us fully on a CT trajectory. It is not even certain whether society has made such investments enough for at least the richer amongst us to experience a CT trajectory. Hoping for either CT or a mix of CT and BAU2 where you are among the lucky world citizens seems an awfully risky bet to me. But perhaps you are more optimistic, looking at Figure 12 and thinking that we can bet humanity as a whole is following the CT scenario. I would love for that to turn out to be the case. Nothing would make me happier than to be proven completely wrong about any and all of this 20 years from now. I shall be mocked for the rest of my life by writers of books like *Why Environmental Alarmism Hurts Everyone* and die an old, satisfied woman knowing future generations will be fine. But there are several good reasons not to be this complacent. I typically avoid delving into them, but I will do so here.

Firstly, there are World3's limitations. As with any SD model, we cannot draw detailed, quantitative conclusions from it. As the *LtG* authors (Meadows et al. 2004) were careful to point out, this is especially true for the collapse pattern because the interactions among variables in World3 are bound to change once the decline has set in. So, even if global society followed CT, this would not necessarily mean declines could be assumed to be as moderate as they are in Figures 12 and 28. My results indicate that global society might experience a halt in growth in the medium term because this is what happens in both scenarios. We can expect declines in CT to be less dramatic than under BAU2, but we cannot be more precise than that. Even if I trusted the world is mostly following CT, I could not assume that the declines would be as small as in the scenario graphs. In theory, this allows for the possibility that these declines could be even more benign than depicted, but there is a better argument to be made for them to be more severe. Because if anything, World3's simplifications seem mostly biased towards optimism. The absence of distribution is one such major simplification, as already mentioned. The global model works with an average citizen, without a distinction between the rich and the poor. Social inequalities are not in the model either, as World3 is a society without any discrimination, oppression, or violence. There is no food waste in the model, and no military or space travel to take resources away from the productive economy. There are also no wars, no strikes, no corruption or fraud scandals, no nuclear accidents, and no pandemics. Because World3 does not distinguish between geographic parts of the world, local natural disasters like floods or droughts are absent. As the *LtG* authors stated (Meadows et al. 2004, p. 221), all these limitations probably make World3 "wildly optimistic".

Secondly, mine is hardly the only research to point to the risk of ecosystem collapse. Among other studies, the Intergovernmental Panel on Climate Change giving humanity a "code red" in 2021 and researchers concluding that we are already on the brink of five climate tipping points in 2022 (Armstrong McKay et al. 2022), humanity's ecological footprint being above Earth's carrying capacity since 1970 (GFN 2021a), scientists concluding that at least six out of nine critical planetary boundaries have already been crossed (Steffen et al. 2015; Masters 2022), and the global insurer Swiss Re (2020) calculating that one-fifth of countries are already at risk of ecosystem collapse today all support more of a BAU2 trajectory than a CT one.

Thirdly, the assumptions about technological innovation and diffusion rates in the CT scenario are not just unprecedentedly high in the sense that they are slightly above historical figures; they are significantly higher. For example, technological progress rates in CT are assumed to be 4% a year. Amongst other things, this should lead to reductions in pollution emissions of 10% from their 2000 values by 2020 and 48% by 2040. Whether we are looking at CO_2, plastic, or any other chemical pollution, we are not on track for such reductions by 2040, and we know for sure that such 10% reductions did not happen by 2020. The *LtG* authors mentioned in their book that they set these rates this high because otherwis, they would obtain a collapse pattern, which would then make the CT scenario not different enough from the other published runs. As mentioned, this more optimistic CT scenario was added to address the criticism that the *LtG* authors underestimated the power of technological innovation, so they erred on the side of overstating it. In their last LtG book, however, the authors also share that it took effort to find this scenario because most (unpublished) runs of World3 indicated that as long as growth continues, new limits will be met. Notably, these limits came faster each time and became harder to innovate out of, especially once at some point they start to occur simultaneously. As the authors put it (Meadows et al. 2004): "the more successfully society puts off its limits through economic and technical adaptations, the more likely it is to run into several of them at the same time. In most World3 runs, including many we have not shown here, the world system does not totally run out of land or food or resources or pollution absorption capability. What it runs out of is its *ability to cope.*"

This is a key but often missed aspect of *The Limits to Growth*: the 's" in "limits". Plural. Humanity can innovate itself out of one limit, but like systems thinking teaches us, this will come with side effects that are bound to trigger a new limit sooner or later. For example, the reason why fossil fuels turned out to be more abundant than feared in the 1970s is indeed that we innovated ourselves, to some extent at least, out of this limit by developing technologies that made it possible to extract deeper and more dispersed resources (e.g., Helm 2011; Faucon 2013; The University of Texas at Austin 2019). But soon a new limit emerged, in this case the pollution that results from extracting and using fossil fuels (e.g., Woody 2013; Jakob and Hilaire 2015). I should point out that my research did not show a close alignment between the pollution variable and the empirical CO_2 data. But the truth is that CO_2 is not a very suitable proxy. The combination of damaging effects from climate change is much higher than the impact pollution is modeled to have in World3 (both versions). In BAU2, pollution levels literally go off the scale (Figure 12), but it is well-established that increases in CO_2 levels much smaller than those depicted in the BAU2 graph would cause ecosystem breakdown (e.g., Intergovernmental Panel on Climate Change 2022). At the current impact factor in World3, other forms of pollution would likely have been a better approximation. Much localized chemical pollution, e.g., water, land, and air contamination, by now has a persistent occurrence in locations around the globe. However, as I will lament at more length later in this chapter, no global data repository of any kind exists for these contaminations. I used CO_2 because greenhouse gases were the only globally polluting substances adequately tracked at the time of my research. So, if we look at the CO_2 proxy (the plastics proxy showed a closer fit), in this proxy as well, World3 is probably too optimistic to be complacent. As I mentioned in my previous point, by now, an overwhelming abundance of research demonstrates that pollution today, notably but not only in the form of greenhouse gases, has become the new constraint to carrying on business as usual. And we are fast running out of time to cope with it. As the *LtG* authors put it in their third book, given enough time, humanity can probably find solutions for just about everything. Time has now become our scarcest resource, however. That's the thing with exponential growth; at some point, it always grows faster than you saw coming.

But the technologist could argue that technological developments are increasingly geared toward sustainability and that the power combination of electrification and solar photovoltaic energy generation is about to completely change our trajectory. I could make the counterargument that, as long as a system's goal has not changed, innovation and technological development within that system will serve that goal, not solve the problems created by that goal itself. The technologist could point at the increases in renewable energy investments and say "See, it's already starting!" I would argue that these investments constitute incremental change in a system that needs a transformation. He could counter: "No, no, it's a tipping point! You mentioned those yourself". We could keep going back and forth with our arguments until we have completely run out of time. That is why I prefer to avoid this discussion. It narrows our choice down to whether or not we should try to avoid collapse. But as Eisenstein (2018) already put it in his 2018 book *Climate: A New Story*, if impending doom was enough motivation for humanity to make the necessary changes, we would have made them by now. Tragically, society's inability to cope with just about everything these days —just as the *Limits to Growth* principle foretells—seems to have ever more people run into the arms of technology gurus, mainstream economists, and conspiracy theorists who promise that some force, be it human ingenuity, the invisible hand, or some day of reckoning, will solve our systemic problems for us (Vargish 1980). Instead of answering the question of why I think CT is unrealistic, a much more pertinent question arises. It's typically not considered polite to answer a question with a question, but in this case, it's the right answer: Do we want to be following the CT scenario in the first place? Why pray to tech to save us from our sins when we can also stop our bad habits and ensure a safe society ourselves? Humankind can now manipulate life at the DNA level, fly into space, and change global weather patterns. With this global reach and unparalleled power to shape our own destiny, we get to—and must—make a conscious decision about who we want to be, and what world we want to live in.

6.4. We Might Actually Want to Save Each Other

> "The law of diminishing returns. This is how they actually explain it. They're like "Paul, do you know how if you have one piece of cake, the second piece isn't as good?" I said: "No, I have no idea what you're talking about. You're eating the wrong cake".

This joke by comedian Paul Morrissey (Dry Bar Comedy 2018) is funny because of course, we know exactly what the person was trying to explain to him. More starts to mean less when we already have enough.

The concept of "enough" was explicitly mentioned in the last *LtG* book. The growth imperative makes sense when we are still working on having our basic human needs met. Shelter, food, and mobility all take material capital. There is a clear correlation between a rise in income and self-reported happiness when a person is close to a subsistence level of income. But as income rises beyond a certain point, its correlation with happiness becomes a whole lot fuzzier. National income level and citizens' level of happiness illustrate this well. Our World in Data (2021) has an up-to-date chart that plots GDP per capita against self-reported life satisfaction for countries. The chart has two scales, linear and loglinear. The default setting is loglinear, and this chart shows a clear upward trend, which seems to confirm a correlation between life satisfaction and income. But that is only because, in this setting, the first increase is from USD 1000 to USD 2000 GDP per capita, i.e., an increase of USD 1000. The last increase is USD 50,000—50 times more than the first increase. Switch the setting from loglinear to linear, and the correlation is gone. If you download the data, as I did, you can fit them. You don't need sophisticated software for this; it is easily done in Excel with "trendline" (Figure 29). If you try out the different options, you will find that the loglinear fit shows a leveling off after about USD 50,000 GDP per capita. The best fit, as indicated by the common measure R^2, is actually obtained through a polynomial fit, which shows a slight downward trend at the highest incomes. Although the difference in R^2 with the loglinear fit is not large, the polynomial fit showing the highest R^2 does suggest the possibility that higher income levels could at some point even start to generate negative returns in happiness.

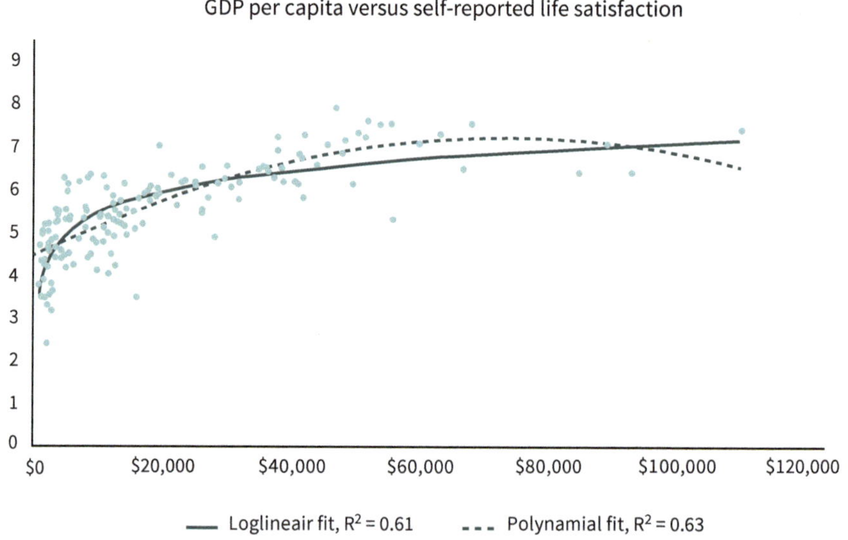

Figure 29. GDP per capita versus self-reported life satisfaction. Source: Reprinted from Our World in Data (2021).

Economic growth not automatically delivering more happiness is not news. Economists have been studying this phenomenon for a while, they call it the Easterlin paradox. The possibility of negative returns after a certain national income level was also already pointed out by other researchers years ago (e.g., Roca 2011). Research indicates that factors that influence the extent to which higher average income translates to happiness include IWI (e.g., Oishi and Kesebir 2016). Lower inequality makes a rise in average income more effective in also raising reported levels of happiness, which, after the section on IWI in this book, probably won't surprise you. Now theoretically, we could decide to just distribute income and even wealth much more evenly while we keep striving for continuous growth. I don't believe this is possible in practice because as I will discuss in the section on narratives in Chapter 7, as long as a society's goal is continuous growth, inequality will always arise sooner or later. But for the sake of argument, let's assume we pair the pursuit of growth with high income taxes on top salaries, a global corporate tax, high estate taxes, and a universal shared dividend of the commons, among other things. This still would not change the dynamic of diminishing returns. Even if we are all seeing a completely equal increase in income over time, we would still need many multiples of an increase in money on the right end of the above chart as on the left for the same increase in life satisfaction. This is a very resource-intensive way to make us happier at that stage, no matter how equally distributed the dividends of growth are. Studies on the micro-level show a similar flattening of satisfaction after a certain income. Some of this confirmatory research shows almost no increase in happiness above USD 75,000 (e.g., Jebb et al. 2018). This flatline is not universal; one recent study, for example, found that there is still a linear increase in happiness with logarithmic increases in salary discernible in some datasets above the USD 75,000 cut-off point (Killingsworth 2021). But note how it is still a logarithmic increase. That inflection point after which more money shows rapidly diminishing returns never goes away.

This begs the question: Is there a different way to give us more life satisfaction that is cheaper? Recent research into well-being suggests there is: helping others. Improving the lives of others improves the well-being of everyone, including the ones doing the giving (e.g., Jackson 2005). This personal improvement in well-being is further boosted by the resulting increase in "social capital" that arises on a macro-level from an improved sense of community and trust, which is observable across world regions (Knight and Rosa 2011). Leveraging such individual and community social capital already seems a much cheaper way to make everyone happier, but there is more. The same policies that improve people's sense of well-being can also carry the additional benefit of reducing our environmental footprint. This is called the "double dividend" (Manno 2002), and it occurs when policies are specifically designed to meet human needs through social relationships and community ties, rather than through more commodities.

My interpretation of all this is that the inflection point, where the relatively steep rise from the beginning starts to quickly level off to a very light incline, is the area where our physical needs have been met, and more stuff starts to mostly satisfy our wants. If you are confused about the difference between wants and needs, that's understandable. The perception that the two are not qualitatively different is a distorted commercialized version of the liberal notion that people have a right to create their own path and are capable of determining the one that is right for them. Of course, I am positively in favor of people determining their own path. Nevertheless, that's hardly the same as elevating the ability to choose between 26 differently scented shampoos, all containing palm oil, to some form of freedom. Has any retail email you received with the subject "This Summer's Essentials" ever been about products that were absolutely necessary for you to have? Of course not.

Needs and wants are actually quite different. One key difference between needs and wants is that needs are universal, whereas wants can vary greatly. Certainly, the subject of which policies work best to satisfy human needs, and in the face of budget constraints, which policies deserve priority, will always be part of the public debate. But human needs are not nearly as muddily defined as some might think. Maslow (1943) did an excellent job already eight decades ago, and his theory resonated with people enough for his name to have become mainstream. Maslow's pyramid (Figure 30) accords with more recent research too. The New Economics Foundation synthesized this body of research into five key activities to promote well-being after our more basic needs have been met (Aked et al. 2008). The first two I already mentioned: giving to others and having strong social relationships. The other three are: being physically active, learning new things, and becoming more aware of our surroundings (the new age term is "mindfulness").

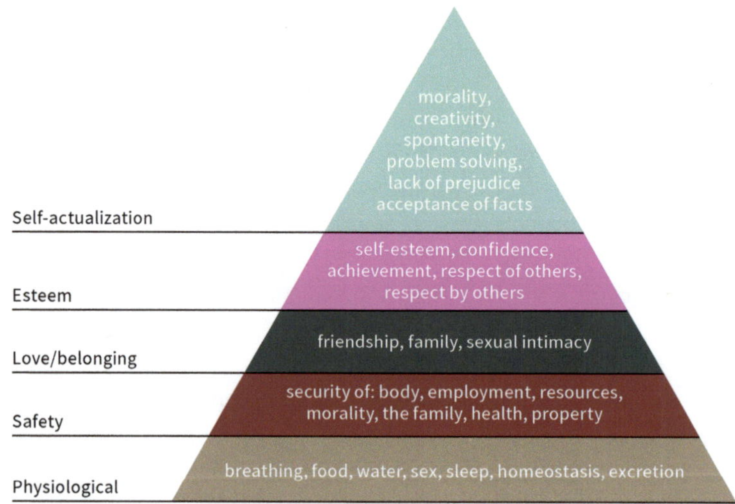

Figure 30. Maslow's pyramid of needs. Source: Figure by author, adapted from Maslow (1943).

Another important difference is that contrary to wants, needs can be met. Once they are, we move on to the next level of Maslow's pyramid. This is not good news if a society's goal is economic growth, because the higher we get in that pyramid, the less those needs can be satisfied with material stuff. Take another look at those last three activities for well-being: being physically active, learning new things, and being aware of our surroundings. They may not necessarily be a social activity (although they often are, of course), but they're also hardly GDP-spurring. Going for a walk in a beautiful park and bringing a library book to read on the bench hits all three of those things and thus can be expected to improve our well-being, but it does very little for GDP. Wants, on the other hand, can grow endlessly. This is especially true when some unconscious underlying need is unmet. That is great news when society's goal is growth. Create an association in people's minds between a product and a need like safety, love, and esteem, and a continuous desire for those products will perpetuate. This is of course where a lot of creativity and innovative capabilities are deployed today. Why do we associate romantic love with diamonds? What other expectation can someone have from a USD 135,000 dog whistle than to garner esteem (Joshi 2020)? Nowadays, most marketing is aimed at playing on our vanities and insecurities. "If you want to get attention or sex, buy our deodorant, make-up, protein shake, clothing, operation, etc." (implicit message: You are not worthy of connection right now. But we can fix that for money). "In these uncertain times . . . ." (implicit message: the pandemic has exacerbated existing social and economic inequalities, so there's a good chance you are more anxious than before. We will fix that for money). "Parenting hack: buy this diaper, book, snack, etc." (translation: We know you know there is no such thing as a parenting hack. But just in case we catch you during one of those inevitable moments you're feeling overwhelmed, we are going to pretend we can fix that for money). Examples abound, I don't have to list more because you are bombarded with some form of implicit message that you are not safe or good enough without this service or that product every single day. The father of modern-day marketing is Edward Bernays, a nephew of Sigmund Freud and the author of books such as *The Engineering of Consent* (Bernays and Cutler 1955), in which he discusses how to increase sales by tying a product to our unconscious desires. Bernays may be little-known, but his legacy continues to this day. As Lustig (2018) writes in his book *The Hacking of the American Mind*, many corporations intentionally blend the notions of pleasure and happiness. People who feel connected to their loved ones and embedded in society tend to experience serotonin and oxytocin levels that are steady and high enough to contribute considerably to their happiness, but that doesn't make them great consumers. Pleasure is a different beast, however, because it is relatively more dopamine-fueled. There is nothing wrong with dopamine (or with pleasure!) in and of itself, but its workings in our body do make it much more suited to exploit for consumerism: it is short-lived, can be achieved with substances such as sugar, alcohol, and caffeine, can more often be enjoyed just by oneself, and in extreme cases can lead to addiction.

This is another way in which growth as our ultimate goal is making almost all of us unhappier instead of, as promised, better off. In a world where nothing can ever be enough, you never seem to be enough. In such a world, the abundance of life turns into a source of anxiety rather than a source of gratitude, because all the extra options mostly represent the many things we still did not do or obtain (Americans called this FOMO: fear of missing out). And the system is configured to exacerbate and exploit these feelings, rather than soften them, to keep those with money chasing the next temporary high, leaving many of us ever unsatisfied. That's why actively reducing inequalities, although necessary, is not a sufficient condition for a sustainable society. Why even in the extremely hypothetical scenario in which we also move from our current linear take-make-waste production processes to fully circular ones without changing our growth pursuit, humanity still would not reach social sustainability. Because for us to stay motivated to contribute to that economic growth, we have to remain at the lower levels of Maslow's pyramid. Indeed, in developed economies, a significant share of the adult population seems unable to rise above Maslow's level of social needs. Unhappiness—defined in a Gallup survey as someone feeling anger, sadness, pain, worry and stress—is at a world record high (Schneider Electric 2021). Reported unhappiness is highest in places of instability, including Afghanistan (especially since the Taliban took back control last year). But unhappiness is also widespread in wealthy states, and there the driving factors include a lack of fulfillment at work and sense of community, i.e., the realm of social needs. From 25% to 63% of adults in Europe, the US, and Israel report feeling lonely some, often, or almost all the time (Ortiz-Ospina and Roser 2020). As mentioned in the last chapter, trust levels are at record lows, and as I discuss in the next chapter, many people feel meaning-deprived. This is bad enough for those of us with money and unmet social needs, but on the other side, it is worse because this system holds no incentive to fill even the most basic needs of people without money. These are the true victims in this system; I want to be clear about that. Having one's basic needs for food, shelter, and safety go unmet is worse than sitting in a big empty mansion feeling lonely. But my point is that in a society aimed at endless growth in which citizens are deemed to be little more than producers and consumers, no one gets treated with full respect.

So, as it turns out, the Easterlin paradox is actually not that paradoxical. It only looks like a paradox when viewed from within an economic framework that elevates perpetual material growth as the apex of human happiness. Once you step out of that framework, it is not at all puzzling that once people's needs and a few wants are met, more consumption does not make them much happier. On the contrary, it becomes quite obvious that the pursuit of growth for its own sake will lead to degenerative systems through many environmental and social channels, and that much more effective and much less resource-intensive ways exist to increase happiness after a certain level of affluence. What makes us happy after that point is, as Nobel-Prize-winning economist Esther Duflo puts it "purpose, belonging and dignity" (Chan 2019). From contributing to our family to donating to the Orangutan Survival Foundation in Indonesia, from caring for the local village tree to organizing history-changing social movements, from protecting one's community to conserving the Brazilian rainforest. All these things, and countless others, can provide us with connection and meaning. Growth for its own sake cannot. That's the beckoning of a call to end the pursuit of growth; an opportunity to reconnect with ourselves, each other, and something greater than ourselves. This new pursuit will make us happier as individuals and lead society as a whole to flourish (Routledge and FioRito 2021).

7. Elements of a Dynamic Global Equilibrium

Global society does not have to settle for CT as a best-case scenario. We have another choice, in the shared narrative that would constitute a shift in priorities as assumed in SW: a new story about who we are, what we value, and what gives us purpose. The previous chapter started with a story because if I advocate for humanity to adopt a new narrative, I thought I should practice what I preach. But it is just that: practice. I don't have the delusion that my story will be the new narrative. That one we have to generate together in an iterative fashion. My story is meant as inspiration to prompt others' iterations. So are the suggestions in this chapter for elements of that vision, and the mental models and structural tools that would flow from it.

7.1. Narrative and Goal

"It is the mothers, not the warriors, who create a people and guide their destiny".

This quote from American author Luther Standing Bear (2006, p. 11) illustrates the juxtaposition of two opposing narratives that have guided humanity throughout history. According to historian and systems thinker Riane Eisler (1988), a society could produce quite different structures, rules, and levels of well-being for its citizens depending on the general value system that would prevail within it. I would call this value system a narrative because the values are described in a story about who humans are. Eisler detailed how two opposing narratives have dominated to various degrees: the domination model and the partnership model. She calls these "forms of social organization", which again could also be called narratives because we organize ourselves through a shared fiction. Eisler places human societies on what she calls the partnership–domination continuum.

7.1.1. Partnership and Domination Models

In the domination model, society is held together by means of hierarchies and strict social scripts about a person's expected role in society based on gender, but also class, ethnic background, and religion. It ranks man over woman, man over man, race over race, and religion over religion. In order to maintain the hierarchies, much of the common narrative revolves around explanations of these rankings as "the natural order". Differences between people are qualifying, meaning they are interpreted as indications of superiority or inferiority. This is not sufficient; the societal hierarchy is further maintained through punishments for deviating from one's social script, sometimes with shaming and other social sanctions, but potentially also with physical violence. A domination model practically always leads to stark social and income inequalities within a society, according to Eisler.

The partnership model is based on relatively flat hierarchies, which are maintained through a peaceful transition of power. Differences are not equated to inferiority or superiority, so social and income inequalities are small, including those between the genders. There is a low degree of abuse and violence, as they are not needed to maintain the many hierarchies. Much of the resources are shared to meet everyone's basic needs. This sharing can be facilitated through democratic hierarchies, but this benign form of governance is not used to oppress. Eisler, therefore, labels this governing form "empowering", as opposed to the "disempowering" hierarchies of the domination model. The two models are summarized in Table 4.

Table 4. Summary of domination and partnership model.

The Domination Model	The Partnership Model
Ranking: man over woman, man over man, race over race, and religion over religion. Differences are equated with superiority or inferiority. Authoritarian structures in family and state or tribe, a high degree of abuse and violence to maintain dominant positions.	Egalitarian structures in the family and state or tribe. Democratic hierarchies, empowering rather than disempowering. Acceptance of differences, but not so much of aggression. Gender equity and a low degree of abuse and violence.

Source: Table by author.

Eisler describes the partnership model, with its emphasis on taking care of one another and nonviolent solutions, as based on female values. However, she is careful to point out that this model is not "matriarchal"; that would imply a woman-over-man power structure which would simply be the same domination model executed in reverse. At the structure level of the systems iceberg, a woman-over-man hierarchy differs from a patriarchal one. But at the vision and mental model level, the two approaches are the same because they're both based on the notion that gender difference constitutes a qualitative difference, and envision a society that grants different rights and opportunities accordingly. This is, I believe, why people steeped in a domination mindset call feminism man-hating or accuse gay rights activists of asking for special treatment, or use terms like "reverse racism". In a domination mindset, empowering one group of people can only happen by disempowering another. But I have never read a feminist argue that the vast overrepresentation of men in crime statistics illustrates that they are too emotional to hold leadership positions, and these, therefore, should be left to women who, because of their more abundant genetic information, higher brain-to-body-size ratio, and higher IQ test scores, clearly are the more intelligent sex. I have not heard any LTBTQ activists narrate how the chaos in society can be ended by returning to the natural two-tier order of, say, "breeders and their leaders". Anti-racists are not using the theory of evolution to argue for a society with "the first people" at the top and the "derivative races", those ethnicities who historically developed later, below them. They all could, in the sense that such arguments would be just as arbitrary and illogical as the arguments for white male superiority and heteronormativity. But to me, it's quite clear that social activists are not arguing for a reversal of social hierarchies; they paint a world in which these hierarchies are dissolved altogether. In this sense, I interpret their work as advocating for the partnership model, or at least something closer to it.

7.1.2. The Link between Expansion and the Domination Mindset

In the long run, domination societies are doomed to fail (Eisler 2008). The domination narrative systemically undervalues anything stereotypically associated with femininity, which leads to structural under-investments in vital parts of society such as education, health, and environmental protection. This makes these societies not self-sustaining, meaning they can only survive for as long as they can expand or exploit land, natural resources, and other societies. Thus, societies with a strong domination model cultivated a drive for expansionary growth. The violence in the domination model then is not just necessary to maintain the strict hierarchies. Another reason that violence was a prominent part of the common narrative about who the people (especially men) were is that this society constantly needed to expand its territories. Additionally, the ever-increasing inequalities and rigid organizational structure dampen a society's ability to adapt to environmental changes (either natural or as a result of the society's neglect). Eisler's work, therefore, puts our addiction to growth in a historical perspective. It explains our collective difficulty in tackling climate change. Our global society has been crippled by inequality so much that we cannot come together even to save our own world as we know it. And these inequalities, Eisler's work tells us, are linked to our quest for expansionary growth, precisely as the CoR already intuited in the 1970s. Contrary to what we have been told, we do not require more growth to reduce inequalities. Quite the opposite. The relentless pursuit of growth and inequality both spring from the same mindset. They will always go hand in hand.

Partnership societies, on the other hand, were sustainable in the sense that they could be maintained with their modus operandi (one can imagine a lot of energy is saved by not oppressing one another). Overall, the partnership model leads to more resilient and prosperous societies than the domination model. The key link between environmental sustainability and the social aspect of equality is also why organizations such as Greenpeace, UNEP, and many more often also address gender and other social aspects. They know by now that to achieve their environmental goals, equality is not a "nice to have" but a condicio sine qua non. Eisler's work also accords with a wealth of studies on the impacts of economic (in)equality, some of which were mentioned in the previous chapter. As Danny Dorling puts it in his 2017 book *The Equality Effect*, it "is almost magical" how in more equal countries, human beings are happier and healthier—even those at the top. However, the partnership model does have one Achilles' heel. When you prioritize caring for one another and don't identify much as a "warrior race", you don't divert many resources towards combat training, weapon creation, or other defensive capabilities. This is fine if everyone practices the partnership model. But you only need one domination tribe or society nearby to make things ugly.

7.1.3. Changing Our Narrative

No one culture is on either end of the spectrum, and our society shows traits of both the partnership and domination models. But if we are to avoid collapse, I would say we need to learn from history and take it way down on the domination part. There are clear signs of ranking man over woman, man over man, one race over another, and religion over religion in the world. They are so ubiquitous and obvious that it feels redundant to elaborate much. I could double my reference list by starting with the global gender indices indicating that in no country women and men are treated fully equally. Maybe I could follow with the fact that men intentionally kill about 87,000 women every year for gender-related reasons. Bring in issues like the racial wealth gap, and how it is illegal to be LGBTQ in 70 countries, with 12 of those carrying the death penalty for this. Then, I would still be leaving out the more extreme examples of modern-day "detention camps" for ethnic and religious minorities that include torture in their routines and sometimes even organ harvesting, and mass rapes committed as a weapon of war, including by Russian soldiers during the invasion of Ukraine which at the time of writing is still going on. Or, I could keep it focused on the US and elaborate on the fact that murder is the third-leading cause of death for Native American women, discuss racial and ableist disparities in the prison system and policing, add a BLM example, and maybe a #MeToo reference. Then, I could capitalize on the most recent mass shooting, of which we have a guaranteed supply averaging more than one per day, to point out that 98% of those are perpetuated by men (who more often than not have a history of domestic violence), while the majority of the victims are women and children. Perhaps I would top it off with a personal anecdote of how my midwife asked me, as part of the intake routine, if my husband ever hit me because one in five pregnant American women experiences domestic violence. But do I really have to?

The sustainability revolution requires a mindset shift towards the partnership narrative and mental model. For one reason, history and the latest research show that a partnership-based global society would make most, if not all, people happier than they are today. Growing this happiness forever would not be possible, but we certainly could increase the life satisfaction of people living today. Even if we didn't care about the next generation or were sure that technology was going to swoop in and save us from collapse, I think greater happiness is a good reason in and of itself to switch to the partnership narrative. Second, as I have laid out in the previous section on CT versus BAU2 likelihood, I don't see any good reasons our global community today can be expected to be an exception to the rule that a domination mindset produces unsustainable societies. This is the most important reason to switch the narrative; it is not working for us and expansion into other planetoids is not really an option. Of course, some billionaires seem to think it is, and I thank them for providing me with a contemporary example of the growth mindset. These men are the few at the very top of today's society. They could, say, end world hunger and still remain billionaires (McSweeney and Pourahmadi 2021). But instead, they put their resources towards expanding their territory even further (by the way, if you're afraid they might succeed in escaping to space, read Sim Kern's tweets of 3 July 2021, to feel better).

It is notable that the domination model, in general, has been more common in places where resources are scarce. This aligns with my conclusion that we are at a now-or-never moment to make a narrative and paradigm shift. If we do not manage to move towards a partnership mindset today, this is less likely to happen two decades from now, once unpolluted resources have become scarcer as a result of more depletion and possibly ecosystem collapse. Indeed, in places more vulnerable to climate change, a rise in gender-based violence towards women, girls, and sexual and gender minorities can already be observed (van Daalen et al. 2022).

That said, Eisler's work also offers hope and a way forward, with her conclusion that partnership is the more natural form of organizing. Despite its Achilles' heel, historical evidence indicates that the partnership model has been more common in the past. Course changes and complete turnarounds of social organization are also not unprecedented and go all the way back to our hunter–gatherer times (Graeber and Wengrow 2021). In his most recent book *A Brief History of Equality*, Piketty (2022) concludes that over the centuries, and although this progress has been far from a straight upward line, humanity has indeed been moving toward greater equality. Other recent social studies (e.g., Hare and Woods 2021; Preston 2022) also suggest humanity is not doomed to perpetuate violence and oppression of one another and nature. That we are wired for connection, possess pronounced altruistic instincts, and have an innate sense of fairness. Accordingly, we see environmental activism and social movements all around the world, with people demanding more equality and inclusion, i.e., precisely the kind of partnership model aspects described above. Young environmental activists like Jerome Foster or Xiye Bastida, only 20 years old at the time of writing, are influencing governments and engaging the younger generation in North America, while Leah Namugerwa, Chibeze Ezekiel, Vanessa Nakata, and Elizabeth Wathuti do the same in Africa. Many of today's major movements are inspired by girls and women, like Tarana Burke, Malala Yousafzai, Marielle Franco, and Greta Thunberg. So many people today being open to listening to women, and recognizing and accepting their leadership could be an indication of our overall changing mindset. True, there are plenty of movements in the opposite direction too. Two of the four activists I just mentioned were shot by men, for example, which in the case of black human rights activist Franco, resulted in her death. And societal collapse is not exactly unprecedented in history either. The best-known book about this is probably Diamond's *Collapse*. But this book also contains several examples of societies that were able to avoid collapse by fundamentally changing their goals, values, and norms. The problems humanity is faced with are on a global scale now, so there's ample reason for the global community to come together to work on them. Despite the dispiriting inaction against climate change, there are also recent examples of society doing just that to solve a common threat, such as the now-recovering ozone layer and the recent COVID-19 pandemic. And there are no space invaders that we know of to make things ugly by exploiting the fact that we prioritized taking care of one another.

What it comes down to is that our narrative needs to mature. This is not a new insight, of course. Greta Thunberg scolded world leaders at the UN Climate Action Summit for not being "mature enough" (NPR Staff 2019), and United Kingdom (UK) Prime Minister Boris Johnson mentioned at the 2021 UN General Assembly that humanity needs to "grow up" to tackle climate change (UN 2021a). But let's put this into the right perspective: acting mature and grown-up is joyous. I am not saying it's not hard work, it is. But despite its perk of much more playtime, very few of us look back on our childhood as the best part of our life. And let's not even speak of that self-obsessed insecure adolescent phase. I don't wonder how my baby will ever repay me for the care I provide for her because it's my pleasure to give (if you think that satisfaction arises only from our shared DNA, try telling that to an adoptive parent and see what happens). My daughter does not wonder about it either. When she is not focused on her immediate needs, she is mostly busy with reaching, building strength, and discovering her voice. This makes sense for her level of development. It does not make her inferior to me. She is simply at a stage where the mindset of "look what I can do" is natural. I am at the grown-up stage, where I get to ask "what can I do for you?" The love of a parent for her child is an easy example (although there are too many heartbreaking stories of child abuse and neglect by parents for it to be a trivial one). But it illustrates my point that, for a mature mind, giving is more satisfying than taking. Generosity is a natural privilege. Being responsible, rather than dominating, is true mastery.

As Diamond (2019) describes in *Upheaval: Turning Points for Nations in Crisis*, we need to decide what parts of our identity we need to let go of, what we want to keep and restore, and what new values we have to adopt. This is easier said than done. Escaping old mindsets requires deeply uncomfortable introspection and constant vigilance for habitual patterns. That is a lot of emotional labor. I for one, however, would not mind putting in the work of shedding the part of the human narrative that describes us as necessarily selfish organisms who only engage in altruism as an aberration. I think we could do without the worldview which tells us that today's stupefying class differences, lavishly adorned with sexism, racism, and homophobia, are just the Darwinian result of inherent differences between us. That the violence, which so paradoxically is required to keep all these "natural" hierarchies in place, too, is intrinsically human. I would not mind a definition of human identity that includes what I think is an innate desire for restoration, to leave the world a bit better than we found it. Where a person's marveling at nature, without wondering what price its beauty or stored energy could yield, is considered not our weakness but the engine for human innovation, or indeed, the moment we're closest to the divine. We could do with a worldview that places humans as a part of life, rather than at the center or the top of it. I could get excited about a narrative that idealizes empathetic service and promises we will be taken seriously for who we are (Golüke 2018). I would enjoy living in a societal system that strives for balance among all life forms, equity between people shaped around universal needs, and hierarchies that place the needs of life strictly above requirements for inanimate entities. Like I've said before, for me at least, it would be totally worth the effort.

7.2. Frameworks and Tools

After we have identified our core societal values and found our common narrative, we can reprioritize our lower-level goals and redesign our frameworks and tools. It's impossible to list all the new possibilities here. But there are already some precursor experiments underway, from the theoretical to the practical. Just as our narrative is not fully formed but many people are working on it, so are the tools and frameworks that flow from these attempts. Below, I will discuss some of the ones most related to my research, and to which *LtG* brings a fresh perspective. I start with what a redesign of our economic system and framework could look like and continue with a discussion on how we work: the ways in which we connect and organize ourselves, how we will measure success, and what quantitative and qualitative information we will want to prioritize going forward. I will also make a few observations on what the *LtG* message means for some key sectors.

7.2.1. Economics

I reserved a separate section for economics not because it is such an important topic in my mind but because it clearly is in the minds of many others. To make this point, Samuelson's quote "I don't care who writes a nation's laws if I can write its economics textbooks" is often cited as an illustration. But by itself, the quote doesn't support much. After all, Samuelson was an economist, and experts have a strong tendency to overestimate the importance of their field no matter what it is. A better measure would be how many people outside of the field use the same language to support their arguments. And by that measure, economics can indeed be argued to be the most important science there is. From taxicab drivers to politicians, from Obama's "trying to figure out how we create an economy where everybody's got a fair shot and if you work hard, you can achieve your dreams" (Winfrey 2012) to Trump's "JOBS! JOBS! JOBS!" (Associated Press and Martosko 2019). So, in this sense, Samuelson was right.

The capitalist system, based on the private ownership of the means of production and their operation for profit, will need to be transformed. In Chapter 5, I mentioned American youth being more positive toward socialism than capitalism, and the Japanese bestseller that was inspired by Karl Marx's writings on the environment, but that does not necessarily mean that the future is socialist or communist. I interpret such findings mostly as a sign that the youth is thoroughly displeased with business as usual and longing for something radically different. Maybe we will indeed adopt a different—perhaps even a completely new—economic and political system. However, that may not be necessary. Capitalism has seen transformations in the past, and it could be transformed again (e.g., Klein 2015). There are many proposals for a new form (e.g., Henderson 2020), such as stakeholder capitalism (Schwab 2021), conscious capitalism (Mackey and Sisodia 2013), regenerative capitalism (Capital Institute 2022), the Mission Economy (Mazzucato 2021), or Benevolent Benevolent Capitalism (2022). These forms vary somewhat in the details of how exactly private ownership should be redesigned, specifically in the role of government, including in how far it should be able to intervene in markets, which parts of society that the economy depends on should be designated as "common pool resources", and how these commons should be managed. However, all these more enlightened versions of capitalism are unified around the argument that an economic system should protect and build what has societal value—which is much more than what can be measured by profits. My personal favorite might be Hunter Lovins' *Natural Capitalism* (Lovins et al. 1999), which proposes that next to manufactured and financial capital, we also measure human, social, and natural capital. *Natural Capitalism* offers four pillars to put this into practice: radically more efficient use of resources, learning from and designing after nature, aligning incentives between businesses, workers, and consumers by moving to a service and flow economy (think of product-as-a-service business models, for example), and regenerating the natural capital that ultimately every company depends on through restoration or reinvestments. I cannot be sure, but I think it is still possible to transform the current "greed is good" narrative into one of these more "enough for each" versions of capitalism.

I feel more certain about the fact that the Homo economicus will have no place in our shared narrative about who we are. We can still use it as part of our economic toolbox, taking out the rational agent in situations where its application makes sense. Perhaps paradoxically, I can even imagine these situations being more plentiful in a world that is aligned with an SW scenario because humans unplagued by physical and economic threats to their physiological needs might just behave more rationally, in the sense that they reflect carefully on expected pros and cons over a long horizon than many people under today's living conditions can be expected to. Still, frameworks that are built on the assumption of the one rational agent, like neoclassical economics, can no longer be the prime framework on which we base policy and business strategy.

Economics used to be called "political economy", because it was obvious that economics is not an exact science, like mathematics, but similar to other social sciences, like psychology or politics (Neal and Cameron 2015). Since the early 20th century, the rise of the neoclassical economics school has made many of us forget that. Nonetheless, as already mentioned, the only reason neoclassical economics can flaunt its many impressive-looking mathematical formulas, giving it the allure of exact science, is because it assumes an all-knowing never-satisfied self-interested-only adult persona who does not represent human beings well. In essence, humanity had to be taken out of economics to make it exact, because humans are not exactly exact. Neoclassical economists tend to speak about their school as "economics" as if there are no other schools of economic thought. In fact, even if you were an economics student at one time, like I was, you may have been taught little other than neoclassical economics. But there have been and still are many different schools of economics, all with their own useful applicability. A substantive well-written overview of economics schools for laypeople is Cambridge economist Chang's (2015) book *Economics: The User's Guide*. Suggestions of past and emerging economic frameworks that will allow for the analyses of today's greatest challenges much more adequately than neoclassical economics are continued in Collste's (2020) more academic read on new economic paradigms. Katherine Trebeck and Jeremy Williams (Trebeck and Williams 2019) also wrote a book on this matter with the apt subtitle *Ideas for a Grown-up Economy*.

Behavioral economics would seem a candidate for the dominant framework of the 21st century because of its gaining popularity. Because this school does not center around the Homo economicus but rather around how humans actually behave, it is better equipped to deliver on what those human beings need. However, it does not explicitly challenge the growth imperative, and so I don't believe that it will suffice as the prime economic framework of this century. It also does not need the pursuit of growth and, therefore, certainly can still deliver valuable insights. But despite being very different from neoclassical economics, to me, behavioral economics still seems too incremental an improvement on our current economic thinking to serve as a key component of a sustainability revolution. Let me put it like this: if the "behavioral" puts the humanity part back into economics, what is this "economics" part? Inserting behavior into economics to me seems to implicitly validate the notion of economics as an exact science when it is not.

My definition of (macro-)economics would be: the (meta-)behavior that arises from people trying to have their needs met. Not wants, as in many definitions, including for example Richard Lipsey's popular one which describes economics as (Neal and Cameron 2015) "the study of the use of scarce resources to satisfy unlimited human wants." This difference is important because in a sustainable society, needs should be prioritized over wants. This means that wants are simply no longer important enough to earn a place in the definition of economics. I have no beef with wants; my vision of a sustainable economy is not one where each person lives near subsistence level and even a modest luxury once in a while is punished. But prioritizing human needs over our wants would necessitate the curtailing of wants to some extent. The question of what products serve a real need and which categorically do not, and in which context, will be a source of continuous public debate. And that is my point: our economic framework should be purposefully shaped around this debate, aimed at enhancing it rather than obstructing it with the fallacy that the market is always right. We do not know yet exactly where that discussion would lead us; however, it is obvious that within an economic framework suitable for a society experiencing the post-era of sustainability transformation, some extravagant luxuries existing today will not make economic sense anymore. The "trying" part in my definition also allows for the possibility of failure to meet one's needs even when a "rational" pathway theoretically exists to having them met. Under certain circumstances, which include individual, structural, and systemic factors, people can make decisions that are not optimal. This can potentially have society-wide impacts, and therefore should be standard consideration in any economic assessment. Lastly, my definition of economics also allows for the fact that needs can go unmet despite material abundance. Unlike what seems to be the implicit assumption in many definitions, including the one by Lionel Robbins, who defined economics as (Neal and Cameron 2015) "human behavior as a relationship between ends and scarce means which have alternative uses", scarcity is not the only limiting factor for well-being. Many people today are suffering in some way not because of resource scarcity but due to meaning deprivation. If there is a lack of capital in that latter case, it's not one of physical capital, but of social capital. But social capital can be "produced" together with little to no ecological impact. So there, again, there is no separating the human part from economics. Of course, there are non-human capital stocks and flows in the economy, such as machinery and natural resources, but those only flow because of human actions (human action is not the only force that moves these stocks, obviously, but those situations are studied in other fields, e.g., geology, ecology, or biology). With my definition, the economy is submerged in society, which of course has always been the case (Polanyi 2001), just like society, in turn, is submerged in nature.

7.2.1.1. Doughnut Economics and Well-Being Measures

A framework that decisively removes the pursuit of growth from economics is Professor Kate Raworth's *Doughnut Economics* (Raworth 2017). In her magnum opus, she introduces the doughnut as the space in between two boundaries (Figure 31). The inner boundary comprises our human needs and the outer boundary comprises the planet's limits.

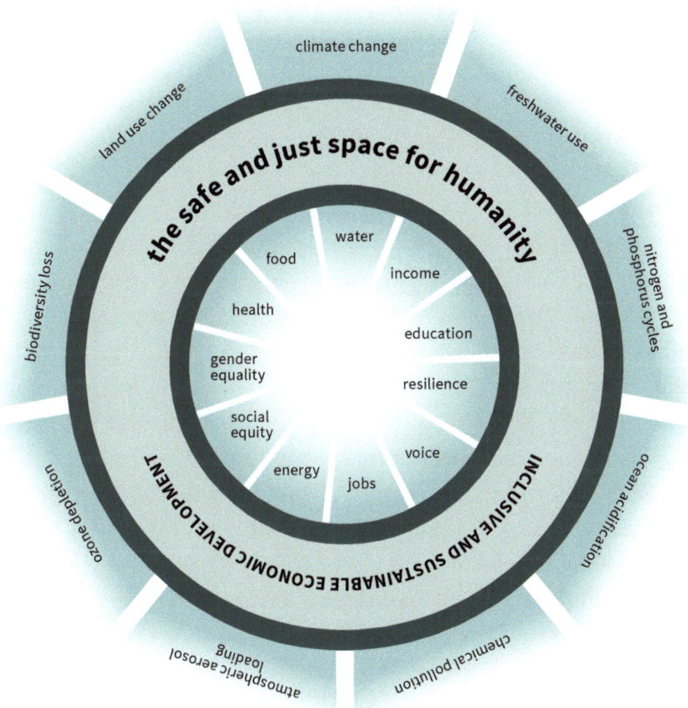

Figure 31. The doughnut. Source: Figure adapted from Raworth (2017).

In this way, the deceivingly simple image of a doughnut introduces the concept of "enough" into our economics. The inner boundary of our human needs represents this concept in the sense of "sufficient". We should aim to have no one fall below this boundary. The outer boundary of planetary limits represents the other interpretation of "enough" as in "no more". We cannot go beyond what our natural environment can carry for us in terms of pollution and extraction.

If you're the spiritual type, you might call these boundaries sacred. Something sacred to us has inherent value; it can never be expressed in numbers. We should honor the boundaries and never cross them, for they hold the sacred space of life. If you're more of the exact type, you could call these boundaries humanity's axioms: Our global society rests on them, and it is illogical to act as if they do not exist because then the entire system comes crashing down. The sweet spot is inside the doughnut: the space of our well-being, that is, where our needs and those of other life forms are met, and we live with the peace of mind knowing that they will continue to be met.

With the doughnut, Raworth places the social and environmental impacts that have far too often been labeled as "externalities" in neoclassical models firmly back into economic practice. This practice is further scaled up in the Doughnut Economics Action Lab (DEAL 2022), which offers a community, tools, and further research for individuals, communities, companies, governments, and other organizations. With the help of the DEAL, for example, the Dutch city of Amsterdam launched its doughnut tool for transformative action in the middle of the pandemic (DEAL et al. 2020).

An economy focused on well-being may sound like a lofty goal, but it is not a dream; it is an ambition. Contrary to perpetual growth, the economic goal of well-being is actually achievable. Practical tools have been developed for years now. The doughnut has been quantified by academia, for example, so we can track countries' performance on it (Allen et al. 2021; O'Neill et al. 2018). And Raworth is hardly the only economist proposing and implementing a framework that puts human needs at its heart. It has been over a decade since the OECD (2021) started the *Better Life Initiative*, which features a series of publications on measuring well-being, as well as the *Better Life Index*. The *Happy Planet Index* (2022), the UN HDI (which I used in my research), or the *Genuine Progress Indicator* (e.g., Kubiszewski 2019) are among the several other alternative measures to GDP. The book *Mismeasuring Our Lives*, published by a group of economists led by Nobel laureate Joseph Stiglitz, French economist Jean-Paul Fitoussi, and Nobel laureate Amartya Sen, has spurred a growing body of research on how to improve the way we measure our economies (Stiglitz et al. 2010). A recent follow-up publication, together with OECD Deputy Director Martine Durand, was the book *Measuring What Counts: The Global Movement for Well-Being*, which is chockful of evidence-based practical guidance (Stiglitz et al. 2019). Speaking of practice, the Wellbeing Economy Alliance (WEAll 2022a) is a collaboration of organizations, movements, and forward-thinking individuals, including economists, from around the world who are sharing well-being economics best practices. The *Beyond GDP* (2022) initiative, which develops "indicators that are as clear and appealing as GDP, but more inclusive of environmental and social aspects of progress" brings together experts from national environmental agencies, universities, and international organizations, including the European Commission (EC), European Parliament, Club of Rome (CoR), OECD, World Wildlife Fund, various UN agencies, and the WB. Some economic thinkers have married well-being more explicitly with concepts such as circularity and biomimicry to form the model of a Nature-Positive Economy: one that is "regenerative, collaborative and where growth is only valued where it contributes to social progress and environmental protection" (UNEP 2021b, p. 18). These new economic models and measures are drops in the mainstream capitalist and economic paradigms, but everything starts small. Their influence is growing, and in this case, that's a good thing. The transformation towards an economy that serves ecological and human well-being might already be underway.

7.2.2. Working Together

If society's goal is no longer perpetual growth, most organizations will need a new goal too. But the way we work in the future well-being economy will be reshaped far beyond just our view of why we work. We will also change our views on who works, i.e., who employees are, and how they are enabled to share their best ideas. The sustainability revolution will require a lot of innovation, but luckily, as it turns out, well-being and innovation come together through new organizational forms at work. We will further change how we work by adopting new ways of communicating and analyzing information. Although incremental compared to the transformations we need, all these changes have been underway for a while now, in some cases, accelerated by the pandemic. The new ways in which we view who works and how to best do so will flow from our views of why we work, so let us start there. Subsequently, I will address the "who" part by discussing the newly emerging social contract between employees and employers. New ways of how we process information in our jobs constitute the "how" part. I will end the section with a few general remarks about the government, tech, and financial sectors.

7.2.2.1. Why We Work

Changing almost everything about how we meet our needs today, and quickly, requires an enormous amount of innovation. We are very much capable of delivering this, but there are two main problems with human ingenuity right now: It is grossly underused and woefully misdirected. Economic and social inequalities leave a large part of humanity's ability to innovate under-recognized, and in some cases completely unused, because this capacity is stored inside the head of a female or a colored or poor person. We've already discussed the importance of allocating more resources to education and reducing inequalities, which in practice would result in vastly broadened access to education, especially for low-income people, minorities, and women. That would address the current underuse of humanity's innovative capabilities but not necessarily the misdirection. Even those of us who were able to develop their cognitive abilities to their full capacity are mostly applying those skills within a society that is economically and politically addicted to growth. If we changed our societal values, most organizations would have to rethink their goals too. In fact, it would be a necessary complement to expanded access to education. Innovation is spurred not only by the knowledge that education provides, but just as much by an "emotional grasp of future opportunities" (Müller 2021). Luckily, humans have this grasp in abundance in the right setting: We are curious, dedicated, and often full of ideas when working together with others towards a shared purpose (e.g., Quinn and Thakor 2019). This purpose cannot be profit. Money will push you to work, but if most of your daily activities feel like senseless tricks for financial treats, you won't be satisfied even if your earnings are high enough for you to qualify as successful. As the 1970s comedian Lily Tomlin put it (People Staff 1977): "The trouble with the rat race is that even if you win, you're still a rat". I think series like *The Office* are popular because, despite their absurdity, the scenes are painfully recognizable. Overall, 25% of the people participating in a cross-temporal survey spanning 47 countries at least doubted the social usefulness of their job. Of those people, 8% were quite sure it was in fact useless (Dur and van Lent 2018). In 1930, economist John Maynard Keynes argued that around now, technological efficiencies would have increased our standards of living eightfold compared with 100 years ago, and as a result, we would not need to work more than three hours a day. Our new problem would be, according to Keynes [1930] (Keynes [1930] 2010), "how to use his freedom from pressing economic cares, how to occupy the leisure". The efficiencies that Keynes foresaw did materialize, and even the eightfold estimate turns out to have been quite accurate (Friedman 2017). So why are we still working much more than a 15 h workweek? David Graeber (2018) argues in his book *Bullshit Jobs*—in which the anthropologist also estimates that about half of current jobs in industrialized economies are useless—that society made a different choice than Keynes thought we would. Instead of reducing the working hours of existing jobs and, as Keynes advised humanity, preparing for our leisurely destiny by "encouraging, and experimenting in, the arts of life as well as the activities of purpose", society kept the workweek the same and invented a whole host of new jobs that our ancestors never could have imagined. Graeber postulates that this choice was the outcome of prioritizing economic growth over well-being. What is a socially useless or useful job is subject to debate. But surveys for quite some time now have indicated that only a minority of employees are engaged in their job. A recent Gallup (2021) survey, for example, found that globally speaking, only about one out of five employees are involved in, enthusiastic about, and committed to their work and workplace. The levels of employee engagement vary by geographic location, but even in the US, where engagement is relatively high, only one out of three American workers are engaged. An important reason for low engagement is a worker-reported lack of trust in corporations, despite plenty of lip service, to strive for genuine improvement in well-being within or outside the workplace. This matters for work satisfaction because most people are motivated by the thought of making a contribution to their communities, and we mentally suffer when we feel like what we do serves no purpose (Dur and van Lent 2018; Googins et al. 2007; Graeber 2018; Pink 2011). But apart from the many millions of unemployed people in the world, even a large share of the gainfully employed don't seem to enjoy the sense of accomplishment that comes from feeling like your work is adding real societal value.

Companies already know this, which is why their mission statements take the form of value creation, a genuine betterment of the world. Of course, the real test is how a company behaves once it seems that profits and the greater good are at odds. These situations are not as frequent as some might think, however, especially not over a medium or long horizon (e.g., Barton et al. 2017). A growing number of companies, including B Corps, demonstrate how doing good often leads to doing well, from startups to giant multinationals, as former Unilever CEO Paul Polman (Polman and Winston 2021) details in his book *Net Positive: How Courageous Companies Thrive by Giving More Than They Take*. Indeed, in this century, companies with a high level of genuine purpose often outperform their competitors (Gartenberg and Serafeim 2019; O'Brien et al. 2019). There are too many examples to go over here, and plenty of excellent books about corporate sustainability already (e.g., Blackburn 2015), including ones listing case studies (e.g., Farver 2019). For present purposes, let me forgo the usual suspects like Interface, Unilever, Patagonia, or Schneider Electric and provide a perhaps lesser-known example from my tiny mother country. The Dutch health care provider Buurtzorg was started in 2009 with the purpose to deliver holistic nursing care. Corporatization of nursing, like in the rest of the health care system and beyond, had put profits at the heart of care delivery. Nurses are allotted specific time intervals for each action, like administering medicine or measuring the heart rate. Buurtzorg's founder decided to adopt a radically different approach which put "zorg" ("care" in Dutch) genuinely at the core of its mission. At Buurtzorg, nurses have complete executive decision on how much time to spend on each patient. Moreover, the nurses also make decisions in small groups regarding management, HR, and operations. There is a CEO, but this person does not serve as the gatekeeper of decisions. Rather, the CEO is available once a nurse group decides, after internal deliberation and consultation with other groups, that they would like this person's advice. No one is checking the nurses' timesheets to determine their individual efficiency. But according to third-party reviews, including those by Ernst & Young and KPMG, Buurtzorg (2022) "has accomplished a 50% reduction in hours of care, improved quality of care and raised work satisfaction for employees". This is because when a nurse takes time to talk and listen to a patient, they can determine what the patient might need in a holistic way, which is often more effective than treating a symptom in isolation under time pressure. Moreover, sometimes, simply that connection is mostly what was needed in that moment for the patient at the time.

Even though companies with a purpose beyond profits sometimes end up being more profitable than their profit-pursuing competitors, they often do not try to outperform other organizations. Buurtzorg's founder is engaging with other healthcare companies, for example, to help them adopt their holistic care model. Houdini produces completely recycled, recyclable, compostable, or renewable outdoor clothing, using special techniques that were often inspired by nature (Martinko 2021). But the company does not guard those techniques; it is open about them, not just for the sake of transparency but because Houdini wants other clothing companies to use these techniques too. Another example is Toyota releasing almost 24,000 patents on electrification and fuel cell technology for other car manufacturers to use royalty-free until 2030 (Tajitsu 2019). Sharing a successful business model or technology with competitors might not make sense in the domination mindset, where an organization's objective should be to maintain a competitive advantage in the pursuit of ever-increasing profits. But it might if a company is genuinely committed to its stated mission, such as delivering quality nursing care or electrifying mobility. We can assume that the motivations of Toyota and that of other companies that released their patents, like Tesla, included the expectation of profitability in the long run. Releasing patents can help jump-start the electric vehicle market, for which they produce products. But nothing is necessarily wrong with benefitting in some ways from knowledge sharing. It is not hypocritical to make profits while on a mission. Just as not pursuing growth as a goal does not mean we are anti-growth, companies do not have to be against profit in order to have a higher purpose than that.

7.2.2.2. Who Works

I once attended a meeting that was booked almost two months in advance with a senior leader who shared with the group of several dozens of people that top management recognized how important innovation was for their company. "So", he said, "what are your innovative ideas?" This is simply not how innovation works, or more accurately, it is not how humans work. As much as we are capable of the radical innovation our society needs, our current organizational and management systems are not optimized for unleashing that capability (e.g., Arekrans et al. 2022). According to Harvard lecturer Leith Sharp (2017), for the innovation that is required for a 21st-century organization, the familiar formal hierarchies will need to be supplemented with non-hierarchical fluid teams. These hierarchies, or command-and-control structures (CCS), are good for scaling solutions—a critical need in the sustainability revolution. But formal hierarchies are not conducive to generating new ideas. Hierarchies are designed to fit the new into the existing structures. In practice, this means that internal procedures require standardized and often quantitative input. But truly innovative ideas almost never conform to those formats, at least not at first. Knowledge of systems, including organizational systems, is practically never completely available in data but stored to some extent in people's mental models. This "tacit knowledge" (Ford and Sterman 1998) consists of a lot of very useful information, but it is not quantifiable. This tacit knowledge can, amongst other things, prompt someone with what in everyday language is sometimes called "intuition". Sharp calls these intuitive responses to observations in the system the organization's "sensing capacity" for emerging issues (2017). We all have this sensing capacity, but if our working environment is not conducive to developing and expressing it, it will not come out. How many other bankers, I wonder, had a feeling something was wrong in early 2007 but could not voice this intuition because it did not appear in (recent) historical data? How many years of rumors and whispered warnings preceded the 2017 MeTooing of powerful men? These are examples of emerging issues that are risks, but on the other side, there are the ones full of opportunity. Sensing these opportunities as they start to emerge is where an organization's real potential for innovation lies (Senge 1994). Almost every company realizes this, of course, but managers and executives often do not know how to put it into practice. Any real move towards innovation would start with a true appreciation for human nature, which would inevitably lead to the realization that a CCS is just too dry and bright an environment for the seeds of our intuition to germinate in. Our intuition is part feeling, and expressing feelings is a vulnerable act. As many researchers and academics have already pointed out, early detection of opportunities for innovation and risks can only be invited in through an environment that is safe (e.g., Edmondson 2018). Safe for us to fail, safe to speak up, safe to be ourselves. How to create this environment is not a mystery, though. The hierarchies we recognize today only become necessary for an organization after a certain group size. Working in smaller flat groups precedes this form. A small unstructured group is a more natural way for us to organize and is also an environment more welcoming to vulnerability. In a corporate setting, Sharp calls such small groups the Adaptive Operating System (AOS). These AOS types of groups have many other names and varying forms, such as "tiger teams" or "pods". But what they have in common is that people in these groups are operating across the organization unencumbered by rank or job description. Roles are fluid because tasks can change quickly, and everyone just does what needs to be done at that moment. Ideas are bounced off each other, picked up and experimented with if they resonate, developed further if they show promise, and quickly abandoned if not. It is a fail-safe setting because the AOS teams are small, and therefore so is the impact of any failure. Personal accountability is decentralized too, delivered to each by anyone in real time. This decentralized continuous feedback is not a trivial aspect of this organizational form. People are social beings and do not need a lot of social cues from their tribesmen—which your team members are to your mammal brain—to change their behavior. This kind of peer feedback is often benign enough, but specific and thus useful. In contrast, annual performance reviews determine your future career and income, and how often have you found that feedback helpful?

AOS teams are not strictly better than the CCS—pedophile rings and terrorist cells work in AOS groups too. The AOS and CCS can be mutually reinforcing, and in combination, they can propel transformations. The AOS is undoubtedly more agile than the CCS, and a more conducive environment to developing new ideas and solutions. The CCS can be used to establish AOS presence with the right policies and procedures. For example, a growing number of companies feed customer or client feedback into the AOS team structure to help spur ideas. Given the volume this feedback can take, collecting and processing it needs to happen in a standardized way. The CCS can (and must) also create psychological safety with its policies, clearly signaling that leadership rewards genuine innovative attempts no matter what the outcome. P&G's internally prestigious "Heroic Failure Award" is an example of this. And once the AOS has enabled an idea to evolve over this human network to the point where it has matured, the CCS can fit it into the existing organization and scale it up. One form of implementation melting the hierarchy with a small flat human network is the Holacracy management system (Robertson 2015). Their model of overlapping circles reimagines management as one that "defines people not by hierarchy and titles, but by roles". According to the Holacracy website (2022), the management system creates organizations that are fast and agile, and that "succeed by pursuing their purpose"—there it is again.

A reappreciation for an organizational form in which people are recognized for what they do instead of what they are called on their business cards also fits with the imperative for employee well-being. Because of course, a well-being economy will have to consist of organizations that prioritize the same for their workers. This does not seem to be the case for most workplaces today, even in the more regulated and richer countries. For example, a 2022 Gallup survey among US workers revealed that only one out of four respondents felt that their employer cared about their well-being (Harter 2022). A majority of Americans who quit their jobs during the 2021 "Great Resignation" cited feeling disrespected at work in a Pew Research Center survey as one of their reasons for leaving (Parker and Menasce Horowitz 2022). As

Holacracy (2022) puts it: "Management as we know it was developed for factories". Management practices of tomorrow will need to be designed for humans, as also laid out in books like *Humanocracy: Creating organizations as Amazing as the People Inside Them* (Hamel and Zanini 2020). This also means, among other things, that companies will want to focus on people's energy, rather than taking their employees' time. With the unstoppable trend of robotization and our urgent need for innovation, focusing on people's time is not just unnecessary; it's counterproductive. Physically exhausted, emotionally disconnected, and/or mentally depleted people are not innovative, no matter how many hours they spend at work. Energy is defined as "the capacity for doing work". It is this capacity that organizations will want to foster, and that requires a more nuanced approach than tracking logged hours.

Our energy comes from a balance of four sources: body, emotions, mind, and spirit (Schwartz et al. 2011). The spiritual part will be addressed with a genuine purpose beyond profit. But this does not guarantee human-centered workplace management; the non-profit sector is known for high burnout rates and overall stress levels (e.g., Kanter and Sherman 2017). No matter how much you love your job, too many hours with too little rest will simply deplete you physically and mentally. We can only sit still or stay awake for so long, and we can only focus our attention for so long; after at most an hour and a half, we need to take a mental break (Thibodeaux 2017). Our sleep and wake rhythm, called the circadian rhythm, consists of a series of 60- to 90-minute cycles. We are simply built with this 24 h internal clock in our brain that regulates cycles of alertness and sleepiness. And for good reason: without this internal clock, "Homo sapiens would not be able to optimize energy expenditure and the internal physiology of the body" (Reddy et al. 2018). Sleep is a key part of our circadian rhythm too, of course, and vital for our energy. Sleep deprivation has devasting effects on every part of our functioning, from our memory to analytic abilities, from cardiovascular health to fertility, from impulse control to our ability to fight off cancer (e.g., Walker 2017). Since the pandemic, 33% of Americans working from home report taking a short nap sometimes during working hours (Morris 2020). Maybe this progress is lasting, in which case their employers would stand to gain significant productivity gains from the boosted alertness. Lower-income jobs typically don't provide an option to work from home, but employers interested in cutting costs might want to consider enabling workplace naps for their low-wage workers too. According to a recent study in India, for example, enabling "the urban poor" to take short afternoon snoozes in the workplace significantly boosted worker productivity and cognition (Bessone et al. 2021). Remote work has become more common since the pandemic, with many employees feeling that the positives outweigh the negatives. Some employers have shown much less enthusiasm, but those have found that demanding people's presence in the office is not as easy as it used to be. When JPMorgan Chase announced a return-to-the-office policy in 2021, for example, it was accompanied by statements from CEO Jamie Dimon such as "everyone is going to be happy with it, and yes, the commute, you know people don't like commuting, but so what" (BuildRemote 2022). Since then, however, the financial giant has announced a loosened version of the policy, after pushback from its employees (Barrabi 2022). Apple is often ranked as one of the most attractive employers, but when its CEO Tim Cook issued a memo saying that workers would be expected to be in the office at least three days per week again, a petition against this decision was quickly launched. AppleTogether, the group of workers from across the company that started the petition, stated that (Da Silva 2022) "those asking for more flexible arrangements have many compelling reasons and circumstances: from disabilities (visible or not); family care; safety, health, and environmental concerns; financial considerations; to just plain being happier and more productive". As consultant PricewaterhouseCoopers (2021) put it in a paper on the future of the American workforce: "business leaders will need to understand what employees really want and create policies and plans that allow for more flexibility and personalization". Some employers are way ahead of this curve. In the UK, the biggest pilot on reducing workweek hours from 40 to 32 without salary reductions is still ongoing at the time of writing (Lockhart 2022). A total of 70 companies and over 3300 employees are participating in the 6-month pilot, which started on 7 June 2022. Bank CEO Ed Siegel said about his company's participation in the pilot: "The 20th-century concept of a five-day working week is no longer the best fit for 21st-century business. We firmly believe that a four-day week with no change to salary or benefits will create a happier workforce and will have an equally positive impact on business productivity, customer experience and our social mission". A survey at the midway point of the pilot revealed that 86% of the participating organizations are planning to keep the four-day schedules. About half of the organizations said that productivity had improved, while almost the entire other half said it had remained the same (Jackson 2022).

And then, of course, there is the emotional energy source, which I believe, in large part, comes down to taking employees seriously for who they are. It is interesting to me how a large part of the solution for organizational agility coincides with treating employees as adults: pushing decision making down the chain of command as far and wide as possible. Notice how the third-party analysts in the Buurtzorg example next to performance indicators also mentioned a higher employee satisfaction. It is motivating to work for an organization that treats you like a trustworthy person who needs no hourly tracking, fully capable of making their own decisions doing what they chose this profession to do in the first place. In his book *Reinventing Organizations*, organization expert and business coach Frederic Laloux (2014) describes many companies around the world, some of them multinationals, that have adopted radically different management practices aimed at employee empowerment. Some of these companies have completely done away with expense declaration; everyone has a corporate credit card. Some companies do not have a formal approval structure, in which case the rule is that you must consult with everyone in the organization you think your decision will impact, but you are not obligated to follow anyone's advice. A researcher I collaborate with in a CoR commission works at the Stockholm Resilience Centre, a research institute in the governance of social–ecological systems. He shared that on his first workday he was told (Gaffney 2022): "This is a flat hierarchy. We trust you to set your own agenda and we check in every six months to see if we are all heading in the same direction."

These practices would make some managers recoil. But it is worth asking yourself what mental model underlies this reaction. These practices are lunacy if people are selfish maximizers who cannot be trusted to make decisions in the best interest of the organization. But if your view is that most people's biggest drive comes from a need to make themselves useful to their organizational tribe and to contribute to something greater than themselves, it makes a lot of sense to not have them waste their time with approval procedures, monthly progress tracking, and expense declarations. The contrast of a hierarchy's function in the domination versus partnership societies shows a clear parallel with the above CCS and AOS discussion. Just like hierarchies are relatively flat and only used to facilitate decision-making in the partnership model, the role of the more hierarchical CCS should be to support the creative solutions-finding process in the AOS and to scale the results. But a CCS structure that is used to control people, rather than connect them, will drain your company from social and human capital at unsustainable levels. (And that is leaving out this company's social and environmental footprint in larger society for the moment.) The pandemic has provided us with some great examples of a partnership mindset versus the domination one in the corporate setting. Some organizations expanded their remote work technology and mental health support services during this time. Other companies have chosen to monitor how long employees' laptops are active and use ID tracking to check compliance with mandatory return-to-work policies. As already mentioned, strict hierarchies often need a lot of controls and incentives to stay in place, and the ones in companies are no different. Most companies work more with carrots than sticks, although this certainly depends on the geographic location of the workplace. But even these carrots, such as financial bonuses for performance on short-term targets, are based on what seems to be a human narrative ill-fitting with the one we need for the 21st century. I am not saying that employee disengagement, together with the rest of the world's problems, will be solved by giving workers in every company a corporate credit card without expense requirements. But by changing our human narrative, we will also change how we view workers and thus how we should manage them. The details of the internal procedures that will flow from these definitions will differ by company.

However, just like global society needs to take it way down on the domination model, organizations will have to wean themselves off of equating management with control over employees.

Innovating towards the sustainability revolution is not just about working together with other people. At least as much, it is about people working together with technology and nature. The latest technologies are explicitly put to use for human welfare in new frameworks like Industry 5.0 (e.g., Renda et al. 2022). As opposed to Industry 4.0, which is mostly about using the latest technologies to improve efficiencies in resource extraction, factories, and supply chains, Industry 5.0 is about "industrial humanization, sustainability and resilience" (Grabowska et al. 2022). The latest technologies are used for circular design, for example, to reduce waste and resource use, and improve functionality. Instead of robots replacing humans to cut labor costs, Industry 5.0 practices include "cobots": robots that work with humans, freeing us from repetitive tasks so we can focus on high-level decisions and coming up with new ideas. At the risk of sounding like a broken record, focusing on well-being does not mean you are sacrificing profits; less waste, less resource use and less boring tasks often also translate to lower costs and higher agility.

When it comes to the importance of people and technology working together with nature for the sustainability revolution, I cannot put it better than biologist Janine Benyus, the co-founder of the Biomimicry Institute (BI 2022a): "When we look at what is truly sustainable, the only real model that has worked over long periods of time is the natural world". Although we must innovate to change our trajectory, we do not have to invent all that much. Nature has already done most of the invention work for us. Nature wastes nothing and typically gives back more than it takes. Any organization that designs a product, process, or some other system, should integrate biomimicry into its design process. The BI's online repository (BI 2022b) is full of nature's circular and regenerative design solutions, just waiting there to be used by smart companies—for free; nature is generous too. There is a lot of attention on the energy transition at the moment, and for very good reason. But although now is undeniably a great time to invest in renewable energy, I personally believe that for investors with a longer horizon there is at least as much potential in pollution-eliminating ecotechnology. As I will mention in the section on pollution data, humans have been flooding our environment and bodies with over 80,000 chemicals that scientists have linked to neurological, developmental, carcinogenic, and other human health damage. And of course, there is the more easily observable waste of discarded products and packaging, including massive islands of plastic. We will want to clean up all of this pollution. Fungi, bacteria, plants, and small animals can help us break down many forms of waste and artificial chemicals that we have created. One of many examples is the already commercially viable *Living Machine*, which uses UV light, bacteria, and plants to treat wastewater at lower costs than conventional methods and to full compliance with regulatory drinking water standards (US Environmental Protection Agency 2002). Indeed, many promising circular solutions we see today are examples of the most powerful combination in this world: nature getting a respectful helping hand from technology to speed things up a bit, fused with human inspiration.

7.2.2.3. How We Work

- Communicating

We use two main forms of communication within organizations today: verbal and numerical communication. Both forms should be broadened in terms of what is being communicated if we are to co-create effective solutions for society's most pressing matters. Let's start with the verbal form.

- Talking

Many of the frontrunning corporations that strive to move beyond mere profit growth have a strong emphasis on communication in their internal policies (Laloux 2014). For people to work together in the way described above, creatively collaborating around a shared purpose in a safe and worker-respecting environment, communication cannot just be the exchange of task-related information; it needs to be connecting. Corporations have shied away from discussing issues like politics, race, gender, and sexual orientation at work because they are sensitive topics. But they are sensitive because they go to the heart of who we are, and how our society functions. It is not possible to be truly engaged and come up with innovative ideas that address real needs when we feel that part of who we are and what we long for in society is not allowed in the workplace. This is true for organizations in general, of course, not just corporations. I briefly volunteered for the Union of Concerned Scientists (UCS) in a small team of people in California who cared about water equity. After the introductions, the team leader who worked at UCS addressed me and said "I am so glad someone like you is joining us too, Gaya." I don't know what made her say that. My best guess is the fact that I was working in the corporate sector, while everyone else was in academia. Or, it might have been that I was the only one who didn't have a PhD. I was also the only part Asian person, while the rest seemed white, although I don't think it was that. There was something in the mind of the team leader, however, and that is my point. I wasn't offended, but I was surprised to encounter "othering" behavior like this at an organization that was originally founded to, amongst other things "devise means for turning research applications (...) toward the solution of pressing environmental and social problems" (UCS 2022). Soon after that Zoom call, a resignation letter was posted on the internet by a black UCS employee with the title *An Open Letter to the Union of Concerned Scientists: On Black Death, Black Silencing, and Black Fugitivity* (Tyson 2020). I won't go into the 17 pages, but as you can tell from the title, Ms. Tyson was not impressed with UCS' diversity performance. To be clear, I have no strong opinion of the UCS. I left that volunteer group shortly thereafter, so I have almost no direct experience with them, and although I don't have a PhD (yet), I do understand that one observation is too few from which to draw conclusions. I know that the UCS holds a wealth of knowledge rooted in science and has issue stances I agree with, including taxing carbon polluters and reforestation to counteract climate change. I gave this example to illustrate that scientists, environmentalists, and social activists are not immune to cause-fundamentalism. It is easier to see such myopia in corporations because their cause of profit is so obviously empty. But they don't have a monopoly on it. When environmentalist Hardin (1968) argued for a limit on births as the only solution to environmental pressures in his classic paper *The Tragedy of the Commons*, he revealed not just ignorance of the long history of sustainable practices with which humans have managed the commons (Ostrom 1990) but blindness to the power of gender equality to bring down the birth rate without any force. One time, a climate activist told me that homelessness is not an important issue because if we do not fight climate change with all our might, now the entire planet will become uninhabitable. But you cannot talk about climate change without talking about climate justice. It doesn't make sense to work towards solutions with a significant portion of your workforce's brain capacity and potential for alliances in the local community largely underutilized. You can't fix everything at once, so it is true that an organization needs to choose a specific purpose. But any organization, private, government, or non-profit, that wants their employees to be engaged around this purpose will have to mainstream, externally and internally, social issues such as racism, sexism, and homophobia, to name a few. Otherwise, it will not be successful in its mission no matter what it is.

So, what then is a good way to have meaningful conversations around these sensitive issues? Again, it comes down to human needs. That is where we can connect because we recognize them, so we can start there with creating empathy. If this sounds wishy-washy, you should know that there is solid scientific support for the calming effects of empathy. The physiological effects that are measurable in our nervous system when our emotional expressions are met with an empathetic response are well-documented (e.g., Hare and Woods 2021; Porges 2017). This social neurological response is the foundation on which Non-violent Communication (NVC) is built. NVC was created by clinical psychologist Marshall Rosenberg (2012) in the 1960s and 1970s. It is very much not a technique to end disagreements or move parties towards a compromise. I believe that attitude, too, is a mild form of domination mindset; we become nervous from disagreements, so we try to convince arguing parties to give and take a little on the things they want and hope that makes it go away. NVC is almost the opposite in that sense. It is about holding space for everyone's feelings and then allowing whatever unfolds from that. NVC comes from a different mindset in that we trust what comes with people sharing in empathy. There is no coercion or "incentive creation". Participants are encouraged to express feelings and needs, but they don't have to. When they do, they can include requests of the other party, but framed again from their needs. The listeners do not have to try to see things from the other's perspective, and they do not have to honor any request. But you'd be surprised how your perspective can change when you connect with others through our needs. The significance of what you thought you truly wanted can fall away in an instant. NVC has since been used in violent conflict settings in Africa and the Middle East, including the Palestinian–Israeli conflict, Europe, tense situations between authorities and students or the general population around race issues in the US, and many more situations including in both domestic and corporate settings. Examples, including verbatim transcripts of these situations, can be found in Rosenberg's books. More information can also be found on the many websites of NVC centers around the world. As bestselling author and researcher Brené Brown (2012) will tell you, empathy is the best antidote to shame. I believe this is relevant because there's a lot of shaming around our needs. This is why framing criticism through our needs takes coaching or at least some practice for most people. Even knowing all that I have written here so far, for example, I still catch myself apologizing to my friends for scheduling meetups around my weekend naps. I have to remind myself that prioritizing sleep is not lazy but responsible (and also, that I am still enough if I am lazy sometimes). Because I too have grown up in a culture where our need for restoration is equated with being weak. But our needs only make us vulnerable, and tending to them makes us indomitable.

- Data

In combination with genuine communication, i.e., qualitative information exchange, data can provide useful insights. Data availability has exploded in this century. According to former Google CEO Eric Schmidt, in 2010, we generated more data in two days than all the data that were created up to 2003 (Carlson 2010). That's more than a 182-fold increase in just seven years. The digital world has only grown since 2010, and is widely assumed to keep growing (e.g., Schneider Electric 2021). It is very telling then, what type of crucial data are still not generated. In general, environmental, social, and governance (ESG) information has been scarce. This has started to change, especially with regard to metrics on carbon and diversity, but most recently also things like plastic waste and water use. The European Union (EU) was the first jurisdiction to mandate "non-financial disclosure" in 2013 (EC 2022a). In 2020, ESG disclosures kicked into gear in the US, with private companies including Bloomberg (2022), Salesforce (2022), and Refinitif (2022), amongst others, offering ESG data services. Various US regulators have been proposing mandatory disclosures on climate and other ESG data (e.g., Securities and Exchange Commission 2022). Canada announced that it will require the disclosure of climate data from federally regulated financial institutions starting in 2024 (Chell et al. 2022). China released an ESG disclosure standard recently as well, which was developed specifically for the Chinese market, and so far will be voluntary (Tong Lee 2022). Starting in 2023, the top 1000 companies in India must report on environmental data and policies, and in the future, this requirement is expected to extend to other companies (Manne 2018). At the last WEF (2022b) meeting in Davos, 70 major companies committed to reporting on the *Stakeholder Capitalism Metrics*, a set of universal metrics and disclosures created to promote alignment among the existing ESG frameworks. My hope is that these data will be used to create new, and speed up existing, feedback loops from the actions of organizations to social and environmental consequences. Tech plays an important role here too. Real-time feedback is possible, and of course, already happening in many applications, but not prioritized in the realm of sustainability. If you have ever driven through California or other dry areas, you know what I mean. Ever-dry riverbeds are still shown as blue rivers on Google maps. As Professor Saskia Sassen (2014) asks in her 2014 book *Expulsions: Brutality and Complexity in the Global Economy*, why does the plastic dead zone in the ocean not show up on any maps? Why are the areas of dead land still colored green? We cannot trace back consequences if they are obscured from our view.

7.2.2.4. Pollution Data

Despite the data explosion in general and the more recent, steeply increasing amount of ESG data, what is still missing is numerical information on pollution. We have ESG data on waste and carbon pollution from corporations. But often, they do not disclose the toxicity of their products, or how they contribute to other air and water pollution. If we are following BAU2 even to some extent, we will have to manage pollution much better than we are currently doing. But what is not measured is rarely managed. Pollution was one of the two variables most difficult to approximate in my research. The other was natural resources, but that's because it simply is hard to measure exactly what is below the ground. People are certainly trying to measure fossil reserves. Not so much with pollution. For this reason, I think that the pollution variable comparison is the weakest one in my research. CO_2 is obviously an underestimation of total pollution levels. Next to the massive plastic pollution that I used as a second proxy, there are chemicals polluting land, water, air, and ultimately our bodies every day. Their volumes have grown to the point that in the aggregate, they now account for one out of every six deaths globally (Fuller et al. 2022).

The impacts of burning fossil fuels, for a start, go far beyond just CO_2 emissions. The air pollutants such as particulate matter (PM) and its precursors are associated with cardiovascular disease and respiratory conditions such as lung cancer, amongst other things. These PM and PM precursors alone are estimated to cost USD 886 billion each year in the US (Goodkind et al. 2019), and that is still leaving out other pollutants such as allergens, or nitrogen oxides, sulphur dioxides, carbon monoxide and volatile organic compounds that together form smog. For China, an estimate of health costs from air pollution was USD 900 billion a year (Farrow et al. 2020). This estimate, made by Greenpeace, seems conservative to me. Coal is the most polluting of the three fossil fuels (gas and oil being the other two), and China consumes roughly half of all coal in the world compared with about 9% in the US (BP 2021). But this estimate is all there is at the moment, because the Chinese government does not closely track these health impacts.

Then there are the other chemicals in our food, drinking water, consumer products, buildings, and more (e.g., Naidu et al. 2021; WHO 2017). You probably know that lead is toxic to humans. Despite this long-established fact, lead is still found everywhere in our environment, especially, but certainly not exclusively, in developing countries. Even so, lead is only one of more than 80,000 chemicals flowing through society, none of which are rigorously tested on safety for human health, let alone ecotoxicity. But they should be, because studies are finding that many of them are potentially harmful to human health as well as the rest of the ecosystem, and their volume keeps increasing by billions of tonnes each year globally. A study conducted last year found PFAS chemicals, officially chemicals from the family of per- and polyfluoroalkyl substances but better known as forever chemicals, in 100% of its breast milk samples (Zheng et al. 2021). Additionally, Bisphenol A, flame retardant, weed killer, and toxic metals such as lead, mercury, arsenic, and cadmium have all been found in both breast milk and formula milk, so there is no way to avoid them for your baby (Lehmann et al. 2018). In fact, more than 200 chemicals are found in placentas these days, some of which are related to cancer, brain, and nervous system damage, birth defects, or pre-term birth (EWG 2005; Singh et al. 2020). It's not just babies of course; according to several studies, some of which are referenced by the Environmental Working Group (EWG 2022) on a page dedicated to the topic, forever chemicals are now in the body of practically every American. These forever chemicals have been linked to reproductive damage, cancer, and immune system harm in very small doses. The positive news of a recently discovered cheap new method to break down forever chemicals (Trang et al. 2022) notwithstanding, real scalable solutions take time to develop and given its global scale there is, as one expert put it, probably not "one single silver bullet to treat PFAS" (Bendix 2022). Some scientists believe that environmental toxins have been causing a decline in sperm count over the past few decades by up to 50% (Mima et al. 2018; Pizzol et al. 2021). In his book *Sicker, Fatter, Poorer*, Dr. Leonardo Trasande (2019) describes what other effects hormone-disrupting chemicals in our houses, air, water, and consumer products are having on us, the gist of which you can guess from the title. The above reasons and more have scientists sounding the alarm bells for a global "potential catastrophic risk to humanity" (Naidu et al. 2021). Despite scientists' alarming findings, they also conclude in every paper that I have read about this topic that significant data gaps exist. That is why they call for, among other things a "Universal Human Rights protection against poisoning", much more rigorous and comprehensive testing, and data transparency and sharing all the way up to the UN level. This would seem to be the bare minimum. Given the preliminary findings indicating serious potential harm to human health, a good argument is to be made for banning these substances until proven safer than initially suspected. This would be in line with the precautionary principle, formalized in the environmental setting at the UN Rio (UN 1992) Declaration, as noted in the next section. You can tell a lot about a society from what it chooses to collect information on and what not. Clearly, pollution control has not been humanity's priority if we can't even ensure our babies are protected from harmful chemical exposure before they ever enter this world. It is unreasonable and impractical to ask each pregnant individual to do their own research before they buy a product to make sure it does not contain harmful chemicals that can cause damage to their unborn baby. But this is what I had to do using the EWG's database *Skindeep*. So, the absence of a global chemicals registry leaving me without a proper proxy for the pollution variable was hardly the worst of it. Once, I received an email from an intelligent and kind student who had been present at one of my guest lectures, in which I had discussed the above on pollution. He said that he loved how my ideas included the perspective of a mother. I am flattered, but I respectfully disagree with his classification. Each of us is impacted by the neurodevelopmental effects of pollution. Even just focusing on the health damage in babies, it would be a misconception to think this lack of knowledge about environmental pollution should only concern parents. It's easy to see that when the next generation grows up being exposed to chemicals that can cause things like impaired impulse control and lower IQ and health status, the consequences will reach society-wide. It is not motherly to get a grip on these pollutants and chemicals; it's just proper leadership.

- Analyzing Together

I believe that working with humans' sense of values, foresight, and intuition, no matter how imperfect and biased, is an indispensable complement to numerical information flows. I am quite comfortable with numbers but, maybe because of that, I know how deceptive they can be. I have encountered throughout my career what seems to be the strongly held misconception that numbers are neutral and, therefore, more reliable than humans. There is a reason we are advised to include numbers in our resumes, or that most reports consist of pages and pages of them. They exude credibility. I am not saying that numbers cannot be very useful. I did a quantitative data comparison myself, after all, and have based strong conclusions on it. But numbers are never the whole story. Not all data are quantitative information, and not even all the data in the world constitute insight. For insights, we need the perspective and values that humans can bring (at least for now, and speculation on the possibility of robots developing consciousness is definitely outside the scope of this book).

Below, I briefly discuss what I believe to be two general improvements to analysis for this epoch. First is the way in which human foresight could be sharpened. It will not surprise you that I believe systems thinking is the way to do this. Then, I will discuss a few systems tools, i.e., tools that facilitate the extraction of human insight, are equipped to deal with interconnectedness and complexity, and blend the qualitative easily with the quantitative. These tools, then, are different in methodology from mainstream techniques like statistics and econometric modeling. Relying on human insights might feel a bit messy compared with those models—probably because most often, it is. Working with imperfect human beings is rarely smooth (that is not to say: frustrating), and the outcomes are never as precise as with the predictive models we are used to. But in a dynamic world, the precision of mainstream models often comes at the cost of accuracy. Because these models cannot process anything that is not easily quantifiable and neatly linear—in other words, anything messy—they miss a lot of input. The result might just be a precise but inaccurate output. Systems tools help us explore different kinds of possible, internally consistent, futures. Such exploration might aid our understanding just enough to where we can avoid turning the situation from messy to ugly. As mentioned, SD modelers do not presume they can predict in a highly complex world, and I believe this kind of humility in the modeler is not trivial. It is related to the domination versus partnership mindset. Outputs from systems tools are not taken as the final say. They serve as conditional guideposts while everyone keeps a close eye on reality unfolding, ready to pivot when early signs come in that our expectations may need updating. This kind of open-mindedness, our individual mental agility, will be crucial for good decision making in the 21st century, just as much as organizational agility.

- Thinking in Systems

Years ago, when still a financial regulator at The Dutch Central Bank ("De Nederlandsche Bank", or DNB), I attended an interdepartmental meeting. Our colleagues shared with the group that, according to EU law, they would now supervise qualitative aspects, like culture and values, which they would gauge through interviews with leadership, amongst other things. Diversity statistics would also be considered. I said: "That's great. Perhaps over time this would enable us to release some of the harsher mechanistic regulatory requirements". The colleagues who had just shared the news giggled at that. "I don't think other requirements will ever go away. These are just additional things we'll be monitoring from now on going forward", one said. I have focused most of the examples of the growth mindset in this book on the corporate sector. This anecdote is an illustration of what a growth mindset looks like in the regulatory context. Regulators savvy in systems thinking would aim to alter the structure of their area of authority in such a way that the regulator's role is diminished. Obviously, this should not be their prime goal—then they could just close shop and leave the market to regulate itself, and we all know how wrong that can go. But this goal could be paired with the stability regulators are looking for. If anything, the justification for any industry regulation would be that the market is not yet at that maturity level, and the regulator's goal should be to guide the sector towards it. That could certainly mean a growing body of regulations over a long period, especially when the regulated sector is growing, but regulators should be careful not to forget over time what their ultimate goal is. True, this vision of a stable, self-regulating sector might never be accomplished, but that does not mean it is not worth striving for. After all, it is not like regulators have always lived up to their current mission statements. Similar to most financial regulators in 2007, DNB's mission was, and is, to maintain financial stability (DNB 2022). It did not prevent the GFC from necessitating a government bailout of the biggest Dutch banks for EUR 32 billion (Meinema 2018). Just like GDP growth rather than different economic thinking is not always the answer to reducing poverty, more prudential regulation is not necessarily always the solution for financial sustainability. I write about sustainability, but my purpose is not to produce more papers and books. Quite the opposite, my end goal is to make myself obsolete by contributing to the realization of a fully sustainable society. Unfortunately, I feel very job-secure for the foreseeable future, but this does not change my purpose. In fact, the more I think about what constitutes true leadership in any setting, the more I believe it comes down to working oneself into redundancy. But that can only happen when the leader affects real and lasting improvements, which requires them to think in systems.

People have argued, at least since the 1990s, for the adoption of systems thinking in the school curricula (Chen and Stroup 1993). I think that would be great, but we cannot wait for the next generation to grow up and solve our global challenges. We should all become better at systems thinking. Not just environmentalists, world leaders, or decision makers. You and I need to become much better fast at understanding how an action could impact the system before it is taken. First, because, as I mentioned in Chapter 2, good leadership will not be adequately appreciated if too many people are only processing the world at the event level, for they will only recognize action in reaction. The COVID-19 pandemic, for example, has been a very visible event. It has heightened anxiety and exacerbated the many ways in which society's systems have been failing many people. Although the conspiracy theories around vaccines are truly baffling, the underlying sense of institutional failing that primes some people to believe these lunatic stories has validity. But because conspiracy theorists have very poorly developed systems thinking, they focus this intuition on the wrong but visible things like lockdowns, mask mandates, and vaccinations. The second and main reason we need to become way better at systems thinking is that we are all in small and major ways participants in this system. I am pretty sure the people who made that EU law on corporate culture and values were more aware of the influence of mindset on actions than my former DNB colleagues supervising on those criteria (and I know that plenty of other DNB employees are too). Still, just some people at the top being mindful of systems will not suffice for the societal transformation we need. We are out of time for "learning on the job" when it comes to our ecosystem. The adage that "you'll regret the actions you did not take much more than the ones you did take" is outdated. This might have been true when a mistake would lead to feedback in the form of a rejection letter, money lost, or a scar. But now that we have entered the Anthropocene, it is not enough to simply learn by feedback from the system, the so-called "trial-and-error" approach, because we're actually at risk of breaking that system if we err.

When it comes to our impact on the planet, if we are to err in the future, we should err on the side of caution. This is the "precautionary principle" that the UN made explicit in the Rio Declaration of 1992:

"where there are threats of serious or irreversible damage, lack of full scientific certainty shall not be used as a reason for postponing cost-effective measures to prevent environmental degradation".

Of course, we will still act; this entire book is a manifesto for immediate global systemic changes. We need to do a lot of damage control and restoration, or perhaps better called redemption. We will, of course, keep innovating even if we reach a state of global dynamic equilibrium (remember: an equilibrium is not at all necessarily static). But before taking any actions, major or small, we will want to make every effort to understand potential consequences, especially when these could cause harm and suffering. Systems thinking can help us do that. Other things that can help us gain understanding in a world of systems within systems within systems, are systems tools.

- Systems Tools

Thinking in systems is the first step, but putting that into practice requires we supplement the mainstream tools with ones that are more applicable in an environment that is inherently dynamic. Techniques like statistics and econometric modeling will still have use going forward. But their (often implicit) reliance on constancy doesn't make them particularly well-suited to deal with the world's interconnectedness and complexity. Complexity was always part of life, but we are today more ecologically, socially, financially, and technologically interconnected than ever. Consequently, things have only gotten more complex. As we saw in the GFC, among many other events, interconnectedness exposes us to contagion: events triggering one another and spreading through a system quicker than its individual parts can compensate for them. This is not necessarily bad; it offers opportunities for those able to gauge the system's workings enough to make use of its flow-on effects, as I already described in the section on leverage points. But interconnectedness also introduces risks; events can spread through a system that we are part of, yet from a source over which we have little control. As the WEF (2018) warned four years ago:

> "Humanity has become remarkably adept at understanding how to mitigate conventional risks that can be relatively easily isolated and managed with standard risk-management approaches. However, we are much less competent when it comes to dealing with complex risks in the interconnected systems that underpin our world, such as organizations, economies, societies and the environment. There are signs of strain (…), and when risk cascades through a complex system, the danger is not of incremental damage but of "runaway collapse" (…)."

Some systems tools I mentioned in Chapter 2: causal loop diagram (CLD), stock & flow diagram (SFD), and system dynamics (SD) modeling. There are others, such as scenario analysis and wargaming. One tool I often find missing from this list is graph theory, which I usually call network mathematics. The list is still by no means exhaustive, and given the pace of innovation, it will probably be improved upon soon anyway. But these tools do have, and I believe will keep having, one commonality: They are meant to be applied by us working together. The time of the belief in a modeler sitting in a windowless room, building a market-cracking model, is over. Any one expert, no matter how brilliant, will miss crucial elements of the world around them. We know from work like that of Surowiecki (2005) and others that groups of people outperform any one person. Naturally, people come with their personal biases. To cancel these out in the aggregate, an expert group should be as diverse as possible in background, field of expertise, years of experience, gender, and other socioeconomic and cultural traits (Surowiecki 2005). With such a group, the below tools can help extract truly valuable insights.

7.2.2.5. Expert Elicitation

In many situations, the tools mentioned below will start with some form of expert elicitation. Expert elicitation is a broad term for the synthesis of opinions from a group of human expert authorities on an issue that is surrounded by uncertainty. This uncertainty can result from insufficient or a complete lack of data, for example, because it involves a rare or unprecedented event (Ford and

Sterman 1998; Surowiecki 2005). But it can also simply be one of the many real-world situations where uncertainty is a given because the environment is complex enough that the past cannot be assumed to be much indication of the future. Elicitation can be as simple as group brainstorming but is typically more structured in order to optimally tap into people's tacit knowledge, which, as I mentioned earlier, is the wealth of qualitative knowledge stored in the mental models of people that are part of the system. In essence, expert elicitation facilitates the mapping of complex systems because it helps explicate this tacit knowledge. A well-known example of a structured expert elicitation is Delphi, created in the 1950s by the RAND (2022) corporation. The WEF also uses a form of expert elicitation for its risk network map, published in its flagship annual Global Risks report. The participants in the elicitation will have informed themselves with available data, of course. But by themselves, real-time data and the most intelligent adaptive algorithms cannot provide us with a vision or tell us conclusively where to start and focus our efforts. For that, at least today, we still need human insight (McAfee and Brynjolfsson 2012). Of course, by virtue of this tapping into people's sensing capacity for emerging issues, the use of expert elicitation also accords with spurring that much-needed innovation in organizations (Senge 1994).

7.2.2.6. Scenario Analysis

In scenario analysis (SA), possible combinations of future events are created in a few internally consistent alternative storylines. Multiple alternative trend developments of several key determinants are considered, but they are not just randomly put together as in a Monte Carlo simulation, for example. SA is not based on extrapolation of the past and is, therefore, advocated by many organizations as a tool to gain insights into uncertain futures (e.g., UNCTAD 2013; WEF 2017). For example, we do not know exactly how global warming will shape the future, but we do know that its impacts will be major and that we should prepare for them. This is why the Taskforce on Climate-Related Financial Disclosures (TCFD 2017) explicitly recommends SA for climate risk reporting. A scenario can be just a story, i.e., a qualitative output. But there are also ways to make this quantitative, by using an SD model.

7.2.2.7. SD Modeling

As I mentioned in Chapter 2, SD modeling is a way to formalize a system into a model and then create scenario runs with it to learn about the behavior of that system. All the variables in an SD model are endogenous. SD models do not rely on an assumption of equilibrium. They allow for different kinds of variables to be modeled and interact, e.g., social, environmental, financial, and technological ones. There are no externalities (or as I call them, "those things we don't want to be held accountable for"). All these aspects make SD modeling at least a worthy addition to the modeling toolbox of people informing decision makers. However, as far as I can tell, in most organizations, including (inter)governmental agencies, the use of more traditional econometric models is still the norm. With SD modeling, you can also determine leverage points in a system. At the end of this chapter, I included an example of a new SD model for the world from the CoR that helped identify the leverage points in the global system that we need to work in to ignite the sustainability revolution.

7.2.2.8. Network Mathematics

The official term for network mathematics is graph theory. That term does not give laypeople much idea of what it is, so I have come to use both the terms network mathematics and graph theory. It is essentially linear algebra applied to a network. When talking about events that influence each other, network mathematics allows one to map those causal influences. In that sense, it is a formalization of a CLD, like an SD model is of an SFD. With network mathematics, you can calculate some of a network's key characteristics. I will not go into any of the formulas here since there are many books on graph theory and its promise for real-life applications (Amini et al. 2018; Barabási 2003; Newman et al. 2005). But there is one characteristic that I have found particularly useful to work with, enough where I want to highlight it: centrality. Centrality is a measure of how influential an event (or: node) is in a network. Thus, it can indicate the most systemically important single network points. Think of air travel, as an example. We all know there are a few international airports that serve as hubs, points in the network where many flight routes converge. In the network of flight plans, those airports have a high centrality. There are two forms of centrality: in-centrality and out-centrality. A highly out-central node, directly and indirectly, has a lot of influence in the rest of the network. This high out-centrality is very useful for identifying at a specific point in time that system's leverage points, or alternatively, root causes. On the other side, we have in-centrality. The points with high in-centrality reveal where the system is vulnerable. These events are hard to mitigate (when they are risks) or stimulate (when they are desired) because you're working against the flow of the system. Your best bet to reduce vulnerability in those areas would be to work with third parties that have more influence there, as I used to advise clients. But if there are no third parties, like in the global system, our best option, by far, is to be safe rather than sorry.

7.2.3. Government

The above was written more with an eye towards corporations because the lack of meaning that many people experience in their daily work is most obvious there. But the private sector is not the only one gripped by the growth imperative. The above principles of working together in new ways that release our innovative capabilities for a shared purpose will work for every kind of organization. Governments are no exception, but they do have a uniquely important role to play by, in close collaboration with civil society (e.g., Hébert-Dufresne et al. 2022), redesigning policies to make their primary aim the delivery of human and ecological well-being. We will need a strong state for this (Nair 2018). Not an authoritarian one—some seem to equate this term to "strong" in a government context—but indeed a state with authority, by virtue of the support and trust it holds from citizens. As the *Financial Times*, not exactly a left-leaning paper, put it (The Editorial Board 2020):

> "Radical reforms—reversing the prevailing policy direction of the last four decades—will need to be put on the table. Governments will have to accept a more active role in the economy. They must see public services as investments rather than liabilities, and look for ways to make labour markets less insecure. Redistribution will again be on the agenda; the privileges of the elderly and wealthy in question."

Through the UN, world governments committed in 2015 to the Sustainable Development Goals (SDGs): a comprehensive set of 17 goals spanning from an energy transformation to hunger eradication, from wildlife preservation to access to education, from non-violence to trustworthy institutions, and more. But the world is, to use the words of UN Chief António Guterres, "tremendously off track" in achieving the SDGs by the 2030 deadline (UN 2021b). Well-being economists Marcello Hernández-Blanco and Robert Costanza argue in their 2021 paper that achieving the SDGs is only possible by applying the systems lens and making use of the interconnectedness among the SDGs. This point was also made by a team of scientists, including *LtG* author Randers (Randers et al. 2018), based on a global SD model, which they later built out into the Earth4All model. Hernández-Blanco and Costanza (2021) also argue that "the SDGs will only be achieved if humanity chooses a development path focused on thriving in a broad and integrated way, rather than growing material consumption at all costs". It is the same refrain: we will not achieve a sustainable world, defined by the SDGs or otherwise, by holding on to a never-ending pursuit of growth.

Some governments have been making deliberate efforts to put nature-positive and well-being policies in place. One small group of front-running state governments that have been exchanging best practices for policies that go beyond the growth pursuit are calling themselves WeGo. They are part of the WEAll, in a subdivision for governments, founded on the recognition that "development in the 21st century entails delivering human and ecological wellbeing" (WEAll 2022b). The WeGo members—Scotland, New Zealand, Iceland, Wales, and Finland—have been putting the recommended policies of the WEAll *Wellbeing Economy Policy Design Guide* into practice (WEAll 2022c). Of course, it is not necessary to be a WeGo member for a government to show sustainability leadership. Finland is one of the few countries to at least be close to achieving the SDGs, but so are Sweden and Denmark, and they are not in WeGo (Sustainable Development Report 2021). Neither is Bhutan, which next to a Gross Happiness Index also uses a cap on income and wealth inequality as its governing tools, and according to the WB (2022c) has "made tremendous progress in reducing extreme poverty and promoting gender equality". Bolivia adopted a Law of Mother Nature in 2011, granting the natural world rights—such as the right to not be polluted and to continue natural cycles—three years after Ecuador incorporated rights of nature into its constitution (Vidal 2011). Both laws were strongly influenced by the countries' indigenous population and the principle of *Buen Vivir*, which translates to "good way of living" and denotes the concept of living in harmony with other people and nature amidst the fullness of life (Rapid Transition Alliance 2018). The Government of Canada (2021) included a well-being/quality-of-life framework in their 2021 budget, two years after New Zealand did so for the first time. The EU has been conducting quality-of-life surveys for ten years now (EC 2022b). Its European Green Deal raises the bar on a wide range of sustainability issues, including eliminating pollution, climate action, sustainable mobility, biodiversity, sustainable food systems, clean energy, and sustainable industry and buildings (EC 2019). Norway has started to pay countries to not cut down their forests, a strategy that demonstrates global leadership and, more importantly, seems to work (Roopsind et al. 2019). The Costa Rican government is an example of climate leadership, perhaps inspired by its philosophy of *Pura Vida*—which translates to "simple life" or "pure life", with connotations of happiness and relishing life. The country's comprehensive and ambitious net-zero carbon plan was designed to transform almost every facet of its economy and, according to a RAND analysis, the country is actually on track to have its carbon footprint reduced to zero in 30 years (Groves et al. 2020). Although there are up-front costs to Costa Rica's plan, the return on investment of citizens' tax money is estimated to be over 100%, in other words, more than a doubling of the total sum. These countries, among others, demonstrate that a government that works with and listens to all constituents can institute systemic changes that make sense economically, socially, and environmentally. Policy details will depend on regional specifics, which among many other things will also include cultural aspects. But whether we call it well-being, happiness, Buen Vivir, quality of life, Pura Vida, or something else, such policies will flow from what we decide has true value.

One thing is clear, though: redistribution is an indispensable part of the solution. As I will discuss in a later section, one leverage point in the global system is inequality. Governments play a key role in this area. They are not the only power; private organizations make choices on issues like how much leadership is paid compared with the rest of the employees. For example, one Seattle-based company made news

in 2015 when they decided to pay everyone USD 70,000 a year, including their CEO
(Cohen 2015), because this is not exactly the standard. CEO pay has increased by
1322% over the past four decades, with a current average CEO salary that is 351
times that of a typical—not the lowest paid—worker (Mishel and Kandra 2021).
But the government has the unique power to directly set taxes on every income
and additionally on wealth, and divert the tax revenue towards actively reducing
inequality through features such as social security, public education, and universal
health coverage. Additionally, this tax revenue could go towards funding research
in resource efficiency and pollution abatement. Given the current status of most
countries, this would align them more with the SW scenario. It is beyond the scope
of this book to go into taxonomic details; it is also unnecessary because there has
already been plenty written on concrete solutions. An example complementing
the abovementioned works from people like Piketty (2014), Stiglitz (2012), and
WEAll is the recent book *The Triumph of Injustice: How the Rich Dodge Taxes and
How to Make Them Pay* by Emmanuel Saez and Gabriel Zucman (Saez and Zucman
2019). The OECD (2022b) and IMF (Gaspar and Garcia-Escribano 2017) have been
advocating for fiscal policies to tackle inequality for years and have written several
papers and reports about concrete measures. On the more extreme end, we also
find a new political concept that has gained some attention over recent years, which
actively opposes the growth mindset: limitarianism. It sounds somewhat close to
libertarianism but its standpoints are not. According to the philosopher who coined it,
Ingrid Robeyns (2019), limitarianism is the concept that "no one should hold surplus
money, which is defined as the money one has over and above what one needs
for a fully flourishing life". Where this surplus begins is a topic of philosophical
discussion, because it depends on our definition of a fully flourishing life. Once
this threshold is defined, the next step might be to conclude that the government
must act on this total wealth cap by installing a 100% tax above it. Where to put this
cap then has become a political discussion. Wherever the effective tax rate should
fall, higher taxes alone would most likely not be sufficient to regenerate our frayed
social fabric. For example, they would need to be combined with access to quality
education and health services (Banerjee and Duflo 2019). Nevertheless, redistributive
taxing measures are a necessary condition. Such measures are not anti-rich; at worst,
they're anti-poverty, but in reality, they're pro-well-being.

7.2.4. Technology

Many argue that the next revolution will be a technological one. I have no
argument with that. The sustainability revolution requires unprecedented rates of
technological innovation. But without a new narrative to guide it, technology will
not save humanity any more than an anti-sexual assault app saves women from
the systemic problem of men not holding each other accountable for rape culture.

Nothing about a call for a new narrative and change in values conflicts with a technological revolution. Quite the opposite; in fact, many experts in this field are calling for the same. They are pointing out that much of the artificial intelligence (AI) today exhibits the same biases we see in the rest of society. Cathy O'Neil (2016) details in her book *Weapons of Math Destruction* how algorithms not just perpetuate but actually increase systemic inequalities. Her book also pointed out how social media can threaten democracy by eroding our social fabric and underprioritizing our privacy, years before the Facebook whistle-blower Frances Haugen (Duffy 2021) and Twitter whistle-blower Peiter Zatko (Bond and Dillon 2022) sounded the alarm on precisely that. Both whistle-blowers accused their former employers of putting profit before the public good. Internal documents that Haugen released show many employees on internal discussion boards voicing their concerns about how the hate speech and misinformation on Facebook and Instagram were affecting societies around the world. This illustrates humanity's state right now: alarming and yet, with hope. Alarming for obvious reasons; we need to improve this because without a well-functioning democracy and healthy social fabric we will not have the strength or even the will to come together and make the revolutionary sustainability changes that society needs. Misinformation is not new, but it needs to be continuously counteracted and with every new medium this battle flares up again. The hope I see is in the public outrage. I wonder if 30 years ago, the news that a company chose "expansion in new areas" over improving the safety of its users would have received the same level of attention Facebook/Meta enjoyed last year. (At the time of writing, the implications for Twitter of Mr. Zatko's testimony are not yet fully known.) Milton Friedman's (1970) statement that the only social responsibility of a business is "to use its resources and engage in activities designed to increase its profits (. . .)" was the accepted doctrine. There still was no law against this kind of corporate behavior in 2021, so the fact that there was a whistle blown at all is progress. The only question is whether it is sufficient progress and whether we will make the necessary social change quickly enough. (Speaking of slow or fast system change, i.e., working in a low or high leverage point, changing a company's name is a good example of working in a low leverage point.)

This unique moment in history of humankind's adulting is coinciding with an explosion of technological development. But it's not an uncontrolled kind of explosion; it is a rocket launch. And we have a choice in the rocket's trajectory. In many places, it is already more expensive to keep existing coal plants going than to build renewable energy facilities (IRENA 2021); now the question is whether we ramp up investment in "clean coal" or in building those plants. We can alter crops genetically, but will we apply this to make them more water efficient and help eliminate pesticide use, or to patent them to oppress farmers? The Internet of Things enables us to connect and collect more information than ever, we have a choice in whether we let it increase the existing inequalities and add new digital ones by using it mostly to make things easier for the affluent and push the lower class from jobs to gigs, or apply it to vastly improve access to education, health care, and financial services. Algorithms can quickly learn some of our behavior better than our spouse and indeed ourselves, but should they be used to manipulate us or to help us make better decisions for our well-being and strengthen our democracies? Blockchain technology offers secured data recording; we can use it to create bubbles of new high-emission financial assets or to improve transparency and trust in global supply chains and carbon offsetting solutions such as reforestation and wildlife protection (e.g., Rebalance Earth 2022). Obviously, I am only scratching the surface with these examples. Just like we don't know what we will find once the rocket has landed on a new planet, if humanity manages to have the sustainability and technology revolutions coincide, no one today can fully imagine the astonishing possibilities that will unfold. But it would start with our deliberate decision to go on this journey.

7.2.5. Finance

When I was working at the Dutch Central Bank, I represented The Netherlands in the securitization workstream of the Basel Committee for Banking Standards. We were setting capital requirements for potential credit losses. One colleague of mine was working in a workstream on market risk, requiring its own capital to hold as a buffer. Another colleague worked on liquidity risk. And then many more colleagues were working on frameworks for other financial assets. But it was the interconnectedness of these risks that took many by surprise in the GFC. The credit losses in mortgages first only seeped into the securitizations for which those mortgages served as collateral. But the fire sales of securitizations that they spurred resulted in market losses that were multiples of those credit losses. This prompted the need for liquidity, which led to more fire sales, resulting in market losses spreading out into other assets and ultimately causing market liquidity to dry up. This materialized liquidity risk caused significant economic damage, which then led to further credit losses in even more assets, which of course resulted in market losses in these assets, deepening the economic drama. Even financial and non-financial organizations that had never invested in securitizations were brought to the brink of bankruptcy, and plenty tumbled over the edge too. It was a perfect cluster of risks that kept triggering each other, while they were each being managed in isolation.

This is a good illustration of what happens when you try to manage part of a system in isolation, because of course while finance is a tool, the financial system is, well, a system. I am hardly the only one who caught on to this financial interconnectedness. Although it is still not fully integrated into regulatory practices, regulators have published and spoken extensively about it since the GFC (e.g., Bricco and Xu 2019; Federal Reserve Board 2021; Roncoroni et al. 2019). Even researchers outside of the field have written about systems behavior in finance, such as the 2008 *Nature* article titled "Ecology for bankers" by Oxford and Princeton zoologists and biologists (May et al. 2008). As this article points out, and systems thinking generally teaches us, there's only so much you can regulate in a system to begin with. It is not a complicated collection of parts that regulators only have to study to fully understand and subsequently regulate to avoid any future risks. It's complex and never fully predictable. That is why economists (e.g., Chang 2011) have argued that an important part of financial regulation is simply to set limits on how efficient and big finance is allowed to get. I will not go into the tool aspect of finance because I don't think that is the issue holding it back from fulfilling its part in the sustainability revolution. Finance needs to fund the necessary investments, but there is no lack of financial instruments, or of money for that matter, to do so. The issue in finance is how it functions right now as a system. In short, it is too efficient and too big for anyone's good.

A system needs some efficiency to function well, but only in moderation, because it also needs redundancy for resilience. Too much efficiency makes a system fragile; it will not easily recover or fully heal from even relatively minor shocks (e.g., Keister 2010; Stein 2010). A too efficient financial system also weakens the rest of the economy. Finance moves across borders much more easily than the rest of the economy, increasing inequalities domestically because of the higher returns on investment compared to labor, as already mentioned in Chapter 5. Additionally, the foreign investment flow can destabilize developing economies rather than help build them sustainably if it disappears as soon as financial returns elsewhere promise to be higher, which has happened (e.g., Chang 2011; Kose et al. 2003). It also makes it harder to tax, or, easier to avoid taxing. Another way that finance brings fragility to the economy and society, is the by now well-known "too-big-to-fail" conundrum. Today, many corporations qualify as "systemically important financial institutions" (SIFIs), which means they cannot be allowed to go bankrupt because they would drag down the rest of the economy with them. The Financial Stability Board (2021) maintains a list of global SIFIs and then regions, like the EU, and countries maintain their own lists of regional and domestic SIFIs. The existence of institutions that cannot be allowed to go bankrupt brings a "moral hazard" with it. To give a personal example, one day, within the first year of my first job at that securitization company I started my career in, I asked my boss for a meeting. I told him I was unable to analyze the investment products he had asked me to evaluate. I could not run a simulation because there was no information on the underlying data, so I could not assume the distributions we used in our models would apply to these datasets. "To be honest", I said, "I'm not sure how my colleagues do their analyses". My boss seemed a bit amused at my serious tone. He said: "I know you learned the theory in college. Now you will learn the practice". He explained to me that everyone else was investing and so they were too. "What if our competitors are wrong about this investment?", I asked. He said: "What if you're wrong though? See, if I listen to you and we don't invest, we might lose clients to our competitors if you're wrong. But if we don't listen to you and we invest, we don't lose clients even if you're right because all our competitors will have done the same". Such moral hazards create obvious risks, which materialized during the GFC. If SIFIs take huge risks that pay off, they get paid; if they take huge risks that do not pay off, the taxpayers pay because the banks are too big to be allowed to go bankrupt. I have said nothing new here at all; in fact, it's very old news, but that's my point: The too-big-to-fail problem is still there. SIFIs now have higher capital requirements than banks with smaller total asset sizes. And regulators ask for so-called "living wills", which would enable the splitting up of large financial institutions in the event part of them becomes insolvent. But it is unclear whether the regulations are enough, and in the meantime the biggest banks have only gotten bigger (Eavis and Collins 2018). I have not heard many neoliberals point this out, but they should: Markets only function well if the bad-performing companies are removed from the competitive environment. If banks, or internet providers, or any corporation too large to be allowed to go bankrupt, are basically guaranteed by the government, then this is not market capitalism. At best it is big firm capitalism (Jahan and Mahmud 2015). But given the enormous inequality in the US, the situation is reaching dangerously close to oligarchic capitalism, and some have even been tempted to call it corporate communism (e.g., Ratigan 2009).

Finance is a very useful tool but should be kept in its natural place. Money and finance have been around for much longer than capitalism and for most of history were embedded in the wider economy (e.g., Polanyi 2001). The economy, in turn, was embedded in the larger society, which is embedded in the larger ecosystem. We have turned this upside down, as depicted in Figure 32, and it creates instability.

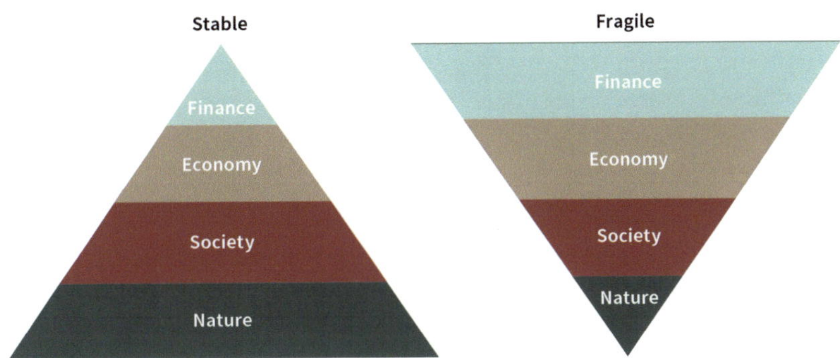

Figure 32. Finance system within ecosystem, stable versus fragile. Source: Figure by author.

There are plenty of policy proposals to "green" finance, which in essence, are all measures to put the financial sector back in line with the rest of the world, which is sometimes inefficient, decaying, and kept from growing too big for its environment. Some of the many examples of such proposals are the Financial Transaction Tax which levies a tiny tax on buying and selling financial products, a legal cap on banking size, or the aspirational but potentially transformative adoption of a global demurrage currency which loses value over time thus disincentivizing money hoarding. But making finance more sustainable does not have to, and cannot only, come from central banks and financial government agencies. Triodos Bank (2022) is a sustainable bank with a mission to make "money work for positive, social, environmental and cultural change". Established in 1980, it weathered the GFC well, and yes, also has been growing steadily with branches in Belgium, Germany, UK, and Spain. The American Aspiration Bank (2022) is a certified B Corp "on a mission to help everyone Do Well and Do Good" by offering financially inclusive products and never investing in fossil fuels. Now, those are small banks. But the CEO of BlackRock, the biggest financial institution in the world, stressed this year that ESG factors are crucial for the giant asset manager, after writing in 2020 that climate change would be a "defining factor" in its investment assessments (Sorkin et al. 2022). In fact, 36% of investment funds are now in so-called ESG funds (Deutsche Bank 2022). All major banks have been publishing about the importance of sustainability issues to wealth creation, with just one of many examples being Deutsche Bank's (2020) podcast titled *Biodiversity loss: how many extinctions add up to economic collapse?*

Are these developments enough? At their historical pace, absolutely not. Although people like me are working hard to change this, terms such as ESG and impact investing today at best mean "less bad" rather than activities having true long-term viability, and ideally, a net-positive impact. Will these developments accelerate at the rate necessary for financing the sustainability revolution? I do not know. But having worked in finance for many years, both in the private sector and later as an international regulator, I dare postulate that the financial sector is one where people are gasping for purpose even more than in many other sectors. Today at least, nothing is sacred in finance. When I say that something is priceless, you immediately understand that I am talking about something that has inherent and inalienable worth, and even trying to price it would be offensive to our sensibilities. In finance, however, value is equated with price, so something that is priceless has no worth. It is hard to find belonging and purpose in such an environment. Indeed, based on my experience, many people working in finance don't like their job. I've had more conversations than I can count with bankers who are open about this but then add the realization that they cannot make the same amount of money anywhere else. Once they have a family, it has become, as one man called it "too late to escape the golden cage". They find purpose in providing for their family and bear a meaning-deprived professional life in their service. My experience is very similar to what journalists have written about their talks with people in finance (e.g., Luyendijk 2015). A recent survey found that junior bankers, who typically do not have a family to provide for, report an intention to quit with a three-out-of-four majority (Moynihan 2022). This would mean then that a wealth of intelligent and ambitious people are locked into a purposeless profession who would love to be part of the solution. And like it or not, we need them to be. We need investment in renewable energy, regenerative agriculture, and developing economies, for example. Some of this investment can come from governments and international development banks, but the private sector is indispensable for the financing of the required sustainability transformation. And let there be no mistake about it: Considering sustainability issues falls squarely within the fiduciary duty of financial players (e.g., UNEP Financial Initiative 2019). If I am right to at least some extent about the financial sector's underutilization of its human potential, it is possible for finance to return to its original function as the economy's blood, moving through the economic body to pick up surpluses here and put them to better use somewhere else. It will not do that as long as it is functioning within an economic framework that has growth as its goal. But by the systemic changes a new narrative will inspire, finance, too, can be transformed into a sector that, as economist Mazzucato (2018) puts it, rather than rewarding value extraction, again invests in value creation.

7.3. *The Big Five Turnarounds: Let's Shoot for the Earth*

Since my research was published, I have had the privilege to work in the Transformational Economics Commission of the CoR. As mentioned earlier, Randers and a team of other systems thinkers have built a new global model, the Earth4All model (Dixson-Declève et al. 2022). It is not the next iteration of World3; it is a new and different model, which has been used to support a new CoR book published in September of 2022, the 50th anniversary year of *The Limits to Growth*. This book, titled *Earth for All: A plan for global wellbeing on a healthy planet*, describes a new human narrative and policy recommendations to bring it to fruition, as part of the CoR's Earth4All project to engage world leaders and policy makers around a 21st century vision for global society. The Earth4All model scenario runs were of course not the only input for the policy recommendations, but they helped bring focus to the thinking that underlies them. Based on the derived insights, the CoR identified five leverage points in the world system. We need these leverage points because, like me, the CoR is convinced that the time for incremental change has passed. Even assuming humanity makes the necessary shift in mindset, we still face the challenge of the enormous speed with which the required societal restructuring should take place. For the sustainability revolution that we need to launch then, we must make use of the synergistic potential in the global system. The CoR phrases this as "breakdown or breakthrough".

The five leverage points are called "turnarounds" in the *Earth for All* book. This term was chosen to signal that these are proposals for a deliberate, drastic break with business as usual. For a while, another term was considered. To quote from an earlier, unpublished, draft of the book: "Rather than ever more testosterone fueled moon-shots, maybe humanity is now ready for a cooperative Earth-shot". I personally like the term Earth-shot better than turnaround, which is why I couldn't resist mentioning it. For consistency, however, I will refer to them as the CoR does. The five turnarounds are Energy, Inequality (Reduction), Poverty (Eradication), Food, and (Female) Empowerment. They are summarized below. But before going into them, I want to point out what they will cost. Because yes, even though in the long run, the turnarounds will avoid many costs—they are less than half the expert estimates of a 10% global GDP loss by 2050 in a "business-as-usual" trajectory from the results of climate change alone (Nair 2021)—they do require upfront investments. The estimated cost of the five turnarounds is 2–4% of global GDP per year. That is a lot of money, but to put this in perspective, the cost of transformative action is less than the IMF estimate of indirect and direct government subsidies to fossil fuels of 7% of global GDP (Parry et al. 2021). Even ignoring relatively straightforward and widely supported changes like higher taxes on the ultra-rich, there is a lot of waste or what could be argued to be suboptimal resource allocation in the current system. For example, about 0.5% of global GDP annually is estimated to be lost due to tax evasion (Tax Justice Network 2022), about 1.1% of GDP goes to the food we throw away uneaten (UNEP 2021a), and 3.3% is spent on just the effects from air pollution that result from fossil fuel use (Farrow et al. 2020). These figures together already get us above the high estimate of the turnaround costs of 4% of global GDP. According to the WB (2022b), gender inequality costs countries twice the global GDP each year. Whatever is holding up the sustainability revolution, it is not a lack of money.

There are many ways towards achieving prosperity, but the analysis by the CoR indicated that these turnarounds are an absolute minimum for the required changes in the global system:

1. **Energy**

    From fossil fuels and energy wastefulness to renewable energy, highly improved efficiency, and electrified transport, heat, and industry;

2. **Inequality**

    From inequality to inclusiveness and fairer distribution of wealth through progressive taxation, trade reunionization, and a universal basic dividend;

3. **Poverty**

From debt and poverty traps in low-income areas to instigating fair and different models for human and planetary prosperity, including new growth models and debt cancellation;

4. **Food**

From extensive, extractive agriculture to regenerative agriculture, diets low in grain-fed meats, and significantly less food waste;

5. **Empowerment**

From discrimination to education, opportunity, and equal leadership participation for women everywhere.

These five turnarounds are, as the *Earth for All* book puts it "our best chance of addressing the destabilising impacts of capitalism and neo-colonialism, stabilising the climate, halting the sixth mass extinction and fairly distributing and using Earth's natural resources". They are depicted in Figure 33.

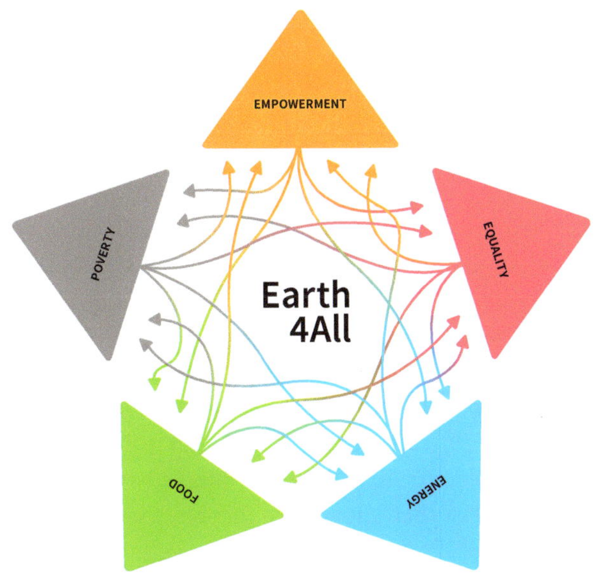

Figure 33. The five turnarounds. Source: Reprinted from CoR (2022).

You can read more in the *Earth for All* book about what these turnarounds involve and how they can be accomplished. But you can already conclude, I assume, that these five areas are deeply interlinked. Food production requires energy, non-renewable energy generation fuels climate change and creates land and water pollution which negatively impact food production, both food and energy impact the larger economic system in which poverty exists, empowering women improves food security and reduces inequality, reduced inequality can positively impact the economy and alleviate poverty, and so on. This makes sense of course, given that these five areas are leverage points in a highly complex system.

The *Earth for All* book is also supported by a website (CoR 2022) and several Deep Dive papers, including ones on region-specific implementations (there is also one Deep Dive paper on my research, which contains nothing that you don't already know if you have read this far). As it turns out, the CoR was not the first to identify the abovementioned five leverage points: The African Development Bank Group (ADBG 2020) had already mentioned them as key focus areas in order for Africa to reach the SDGs. The ADBG's execution details of these focus areas illustrate how the CoR's turnarounds will vary in practice for different regions.

The growth models in turnaround 3, Poverty, cannot be the same methods of growing that the rich countries have used. As I mentioned in Chapter 3, we will not stay within planetary boundaries that way. For low-income countries, economic development that does not degrade natural and social capital while generating greater income is feasible. A commonly held misconception is that there exists a trade-off between the environment and the economy, or at least at the start. This ubiquitous notion claims we must grow the economy at all costs before we reach a point of affluence where we can start to care about inequality and environmental pollution. And many people seem to hold the additional unexamined assumption that once we reach that point of affluence, we will relatively easily and swiftly bring social and environmental costs down. This concept of a reversed U-shape for inequality is called the Kuznets curve. Kuznets only offered it as a potential hypothesis when it seemed to show up for income inequality in the data available in the 1950s. He did so with a bunch of qualifiers about data quality and applicability of the hypothesis, but the economic world just went with it and later extrapolated it onto environmental pollution. (Kuznets had this happen to him before; he is the one who came up with GDP in 1937. Just like with the curve that was named after him, he presented GDP with a few warnings, including that it should not be confused for a measure of a country's well-being or prosperity. But the world just went with that one too!) However, research since then shows that there is no Kuznets's curve observable in empirical data (Stiglitz 2012). Low-income countries can leapfrog to best practices from the start, by growing their economies in an inclusive and environmentally responsible way. This should happen (and already is happening) in partnership with more developed economies, with the latter providing active and consistent support, including financial support. This international cooperation would not be a dynamic of one-way giving and taking. It would not be, as sometimes has happened in the past, a Western country "saving" a developing country with aid that comes with a set of restrictions and demands that transplant Western norms or strengthen Western economic interests (e.g., Nair 2022). It would be an exchange of material, financial, and knowledge resources, with respect for local knowledge and cultural norms. For reasons discussed earlier in this chapter, this kind of exchange can be expected to also deliver purpose, resulting in higher life satisfaction for all parties.

If the word "empowerment" in turnaround 5 rubs you the wrong way, you're not alone. During a panel at the September 2022 book launch event in New York City, panel participant Nadi Albino said to some of the Earth4All team members: "The term female empowerment is dusted and done. You should choose a different term". Per Espen Stoknes responded that he understood her point, and that the team had also discussed this for quite a while. "We decided to keep it", he explained, "because people know what it means and the alternatives are more likely to activate toxic masculinity responses, which are just so unproductive". I think they're both right, and I do not have an alternative wording to propose. That said, I don't particularly like the term either. Female empowerment is easily misunderstood as a call for women to step up, but it really is a call to men to actively counteract cultural and institutional barriers that keep women from sharing with society all the ways in which they are already powerful. Not because this benefits women, but because this benefits everyone. Gender inequality may hold women back, but it also renders men docile under the constant fear of "losing their masculinity". Misogyny is meant to keep men in line with a system that does not benefit most people, men nor women, in exchange for the meagre reward of a self-defeating illusion of superiority (Manne 2018). A superiority, at that, to a group of people which not seldom includes someone they long to connect with the most. Four global leverage points arguably have more to gain from women than men, simply because of women's historic underutilization. But in the global turnaround of female empowerment, it is men who ultimately hold the greatest potential for change.

7.3.1. The Difference between Leverage and Gravity

The five turnarounds of Earth4All do not imply priority in gravity. It is not that we should focus on agriculture because it's more important to feed people than it is to stop today's massive wildlife extinction or aid refugees. Or, that access for women to education is more important than housing security. Weighing one sustainability issue against another, in general, is a losing proposition, no matter your calculus. Leverage points do not indicate how much we should care about certain issues or empathize with those affected by them. It is just that some issues are further down the causal pipeline than others. Because of that, they are not the first levers to pull to accelerate lasting system-wide change.

Water is a good example of a non-turnaround but hugely important area in the global system. Water is like kindness or the female: bearer of life everywhere, and everywhere undervalued. We need water not just to survive but to do pretty much anything. Energy, food, buildings, transportation, and consumer goods take enormous amounts of water to produce and/or use. Yet, we have a finite amount of potable water on Earth. This is not new to people living in drought-stricken places, even the wealthy Southern Californians. Water stress, defined as a condition in which the demand for water is greater than its supply, has been steadily increasing and is projected to keep doing so due to population trends and climate change (UN Water 2021). The UN predicts that within a decade, half the world could face some form of water stress (UNEP 2016). The European Commission estimates a 75% to 90% probability in the next century of wars being fought over water (Farinosi et al. 2018). In short, water is undeniably very important, and we should treat it as a scarce resource, much more than we are currently doing. But it's simply not a leverage point. If I performed a network analysis on the global societal system, I would not be surprised to discover that water is a vulnerable node, as I've described in the previous section on network mathematics. To date, I have not run such a network analysis. But if water were a vulnerable node, it means that once risks materialize there, we have very few options left to act. The problem of water availability would more effectively be tackled by focusing on points in the network more "upstream" in the causal pipeline. The energy and agricultural turnarounds would seem to be points where we have more influence. And indeed, switching to renewable distributed energy generation and regenerative agriculture and reducing food waste would result in direct and enormous water reductions, among many other things.

The five leverage points are full of possibilities for forward-thinking companies. Some non-leverage points, especially those very far down the causal pipeline, like homelessness or shelters for domestic violence victims, in my opinion, will remain the domain of charities and government entities for the foreseeable future, although I would love to be proven wrong on that. These charities and similar community volunteer-based groups will still form a vital part of society in that way. Just because they are not working in a leverage point does not mean they are not addressing a critical need. They are. But the leverage points are just bursting with business opportunities. As someone working in the corporate sector, I like that because I can get others to engage around these issues much more easily. Global transformation requires everyone to play their part, and the corporate sector's active engagement is indispensable. Corporations would still need to collaborate with government and non-profit entities in many situations. For example, the energy transition requires that people currently working in the fossil fuel sector receive social programs that offer re-schooling opportunities and temporary incomes that last the entire transition. And combatting income and wealth inequality obviously requires changes to the tax code. But if you work in the corporate sector and cannot see the energy transition that is already underway, the opportunity at the bottom of the wealth pyramid that financial inclusion offers, the imperative of a healthy middle class for a stable business environment, the underutilized potential in women, or the benefits such as increased land productivity and halting of desertification that the right regenerative agricultural practices can yield, I am not sure you have any business being in business.

7.4. Getting Comfortable with Uncomfortableness

Now that we are getting to the end of this book, I would like to leave some personal advice in case you're planning to contribute to the sustainability revolution in a way that makes sense to you.

First, expect and if possible, accept the tension that arises from working on system change. You must participate in a system in order to change it from within. But of course, you will not change it if you're completely engulfed with its narrative; you will just maintain it. Doing your part in the sustainability revolution means you will have to work in a system the narrative of which you don't fully share. This will bring tension, with others, but mostly within yourself. For just a taste of this on the social front, take some of the association tests of Project Implicit (2022). This international collaborative of researchers has many online tests to measure one's "implicit social cognition", based on things like gender, skin color, ability, weight, and sexuality. When I took the test for gender and science many years ago, it revealed that I have a strong association with beta sciences (natural sciences and math) and being a man. Throughout my life, I have consistently been the best in class when it comes to mathematics. I graduated cum laude in Econometrics. Whenever someone tells me—luckily this happens less and less—that I don't look like someone who's good in math, I ask them "what about my appearance tells you that?" Apparently, according to my own unconscious, it is the fact that I look female. The researchers of Project Implicit will tell you that a test for implicit social association based on gender is not the same as a "sexism test". Your behavior would be the input for the latter test, not your biases. I have a sexist bias. But this does not change the fact that I am a feminist, because every day I challenge this bias, in others and in myself. This is also why anti-racism activists, for example, will tell you to stop focusing on not being racist (DiAngelo 2018; Kendi 2019). We all have racist biases from growing up in a society that tells us that white is better than black or brown. You will achieve more by acknowledging this fact and then making conscious efforts to counteract those biases in yourself and others because that way you are changing the system towards something less racist every day.

Second, take a break sometimes to tend to your needs, or as I prefer to think about it, to work on your happiness. A slew of philosophers will tell you that pursuing happiness is a surefire way to stay unhappy (e.g., Schopenhauer 2004). I used to agree, but I've come to believe that is because most people chase dopamine hits in that pursuit, and that indeed rarely leads to a regular routine in the New Economics Foundation's five activities for well-being like giving to others and consciously appreciating one's surroundings. Ever since I started to plan in time for things like prepping fresh organic meals, walks in the park, and no-screen-connecting time (yes, I put all these things in my calendar), my physical energy and emotional resilience have been strong. I classify this as an achievement, especially given the subject of my research and work field. I realize even those simple activities are not attainable for some people; fresh air, organic food, and living close to greenery these days are luxuries. But even just home cooking, organic or not, will improve the health of your gut (Pollan 2009). This will not just boost your physical health but also your mental health, because your gut health strongly affects your body's ability to produce happiness hormones in the first place (if you're not looking for yet another book to read, watch "How Cooking Can Change Your Life - Michael Pollan" on YouTube). Of course, your mental health deserves direct attention too sometimes. If we are talking about philosophers, I find the work of Albert Camus more useful than any others'. Absurdism seems to hold the best advice for how to carry oneself in a world of constant distractions and "alternative facts". When "standing face to face with the irrational", as Camus (1991) puts it in his essay *The myth of Sisyphus*, longing for happiness and reason, what do you do? It is possible that nothing you do will make a difference. You might lose, and there will be no justice in it, no lesson, no sense. You do it anyway. You commit. I don't want to turn this paragraph into a philosophical treatise—for one because I would be out of my depth—so let me make a popular culture reference. Although probably the only thing that society can agree on right now is that the movie *The Matrix Revolutions* was an abomination, the final fight scene between the main character Neo and his antagonist Agent Smith contains some teachable absurdist elements (Wachowski et al. 2003). Neo is up against overwhelming odds, symbolized by nothing but countless copies of the AI Agent Smith everywhere. After having smacked Neo to the ground several times, and then finally deep into it, Smith is standing in the hole he just made looking down at a floored Neo and saying: "Why? Why do you do it? Why get up? Why do you keep fighting? Are you fighting for something, for more than your survival? […] You must know it by now. You can't win, it's pointless to keep fighting." Neo stands up. Exasperated, Agent Smith asks: "Why, why do you persist?" Neo answers: "Because I choose to". This being a Hollywood movie, Neo is victorious in the end, but that's not the point. My point is the display of the human spirit. I regularly repeat a quote from Gandhi to myself: "Whatever you do in life will be insignificant, but it is very important that you do it". This holds true because you do not know how your actions impact the rest of the system. You may have made no difference, or you might have indirectly changed the course of history in ways you will never realize. Either way, the only choice you ever had was whether to act, the rest is not up to you. It is that high-ambition-low-attachment attitude that Buddhism can also teach us to cultivate. You care, of course; I wish for you to revel in your successes and hope you will allow yourself to be sad in the face of disappointment and defeat. But you'll only be failing when you start to become bitter. When you can't feel that subtle enjoyment anymore from the defiant persisting in your humanity, that's your cue to unplug. Working in systems is overwhelming; if everything is connected, you're never really done. But remember, it also means that you're never working alone. So, when you start to feel like nothing you do will ever be enough, take time to restore and let other parts of the system carry you.

Third, embrace the term "hypocrite" for now. Whatever you choose to focus on, if you're going to make any impact at all, you will run into someone pointing out that your behavior is in some way inconsistent with your proclaimed values, and therefore you are not credible. Take it as a good sign; you are making a change in the world. When I became pregnant, I researched pregnancy and birth books with the same ferociousness as I applied to my *LtG* research. Before that, I had been mostly vegan because I could not justify the disproportionate carbon and water footprint of a meat-based diet, and I wanted no part in the enormous cruelty in the meat industry (in my opinion one of the biggest sins of our times). At one month pregnant, I had to conclude that scientific findings so far indicated that a vegan diet would lead to nutrient deficiencies in my baby. I looked for studies that could support a vegan diet for pregnant women, but the few I found were just not rigorous. This does not mean they won't come out in the future; for ethical reasons, studies on pregnant women's diets are never controlled, and thus it is often difficult to find conclusive findings. But given the research available about the gestational needs of babies, I could no longer justify a vegan diet. So, I ate meat or fish almost every day for months. Was this the right choice? I was doing what I thought was best for an innocent life that I had responsibility for. But of course, the animals I consumed were also innocent and wanted to live, and eating them was my conscious choice for which I thus also carried responsibility. Does it make a difference if I tell you that I ate mostly organic and pasture-raised or wild-caught whenever possible? Or does that just make me privileged? My point is that it is impossible to have one's needs met in this system without causing harm in some way. As I have laid out in this book, a societal system with the goal of continuous growth cannot possibly meet human needs in a sustainable way. That is why we must change the system's goal, and with it the entire system itself. But in the meantime, we still have needs. You do not need to apologize for that. In fact, making people feel guilty about their needs is a tried-and-true tool of oppression and you should recognize it as such. If you make efforts to meet your needs in a way to minimize harm, you are credible. Being this right kind of hypocrite in today's world means not being callous about the unnecessary suffering and destruction in it. As Jiddu Krishnamurti put it "It is no measure of health to be well adjusted to a profoundly sick society." But don't let people use your compassion to shame you into chasing the pretense of clean hands.

The only way to keep your hands clean is to put them to work outside of a messy system, and you will not make any impact there.

Whether your chosen prime cause is to fight systemic economic inequalities, sexism, racism, lack of animal rights, or environmental degradation, to name a few, you will make others uncomfortable, and you will regularly feel uncomfortable with yourself. You will be a hypocrite because you have grown up in a classist, racist, sexist, homophobic, ageist, enbyphobic, speciesist society that systemically undervalues nature in every way. But if you make a genuine conscious effort to be comfortable with your uncomfortableness and be a little less hypocritical every day, you have earned your right to speak up.

Lastly, if you are reading this as a person with significant influence, especially in the private sector, I have some personal advice for you too. I have given this advice before to executives. The first time to an already quite successful tech businessman who told me that the social aspects of algorithms can come later, once he had made a permanent name for himself. Then to the about 40,000 KPMG US employees and partners who are given the firmwide mandatory sustainability crash course for which I was interviewed shortly after I left. And now to you: Do not sell yourself short by underestimating the profoundness of this opportunity you have to be a leader today. Yes, there is an enormous market potential for trailblazing organizations in sustainability. And sure, companies that don't make the financial, physical, and emotional ESG investments won't be around two decades from now. But the possibilities here go beyond your business doing well by doing good. You will have many more opportunities in life to prove yourself a good chief. With this challenge of powering the sustainability revolution, you get to prove yourself a good person. Most people will never have that opportunity. May you realize it.

8. Who, Where, and Why You Are

We have been told that the pursuit of growth is the pursuit of happiness. That for everyone to have a sufficient slice of the pie, we must keep growing it. Fifty years ago, the *LtG* authors warned about the limitations of that narrative. Their work showed the fallacies in the story that continued material growth will keep providing each person an ever-better life. They were criticized, then ridiculed, and over time mostly forgotten.

But my research shows that empirical data today align closely with some of the *LtG* scenarios. This close alignment implies that growth will halt within the next few decades one way or another, and the only choice we have left is its cause: social and environmental breakpoints, or our own conscious action to limit the ecological footprint of how we meet everyone's needs. The two closest aligning scenarios, BAU2 and CT, show that, at best, technological innovation might protect us from a societal collapse but not from declines or a halt in growth. Moreover, most other research aligns more with BAU2, which predicts steep declines in food productivity, industrial output, and population as a result of pollution, including carbon emissions. But there is a better option than to buy a doomsday bunker or pray for technological deliverance. My research also reveals that the SW scenario, in which the highest levels of well-being are maintained throughout the rest of this century, is not yet too far off from empirical data.

This book was written at a historic moment in time. Attention to the *LtG* message has risen again now that more and more people are losing faith in our current systems amidst ever more frequent and intensifying signs that we have pushed beyond social and ecological limits. But half a century later, the time for a paced transition has passed. The transformational changes that are required for realignment onto a path towards a sustainable world can only be made in time by working at the generative level of our collective narrative. We must redefine who, where, and why we are. Ultimately, today's challenges are not about combating climate change or fighting inequality. This is a battle for humanity's soul. It cannot be won by making others lose. And no one can fight that battle other than we. It is up to us. Humans, so self-focused when resources are constrained, but longing for meaning and connection as soon as our physical needs are satisfied. We are open and sensitive, which makes us vulnerable to manipulation, but also capable of cooperating across the globe. Embedded in nature, always part of the web of life in a world the interconnectedness of which we should respect and learn from, rather than try to control. Now that we have reached global power, we have a choice to keep using it for an empty delusion of domination or to direct our might towards genuine happiness. If global society does not make the shift in mindset from a domination model to a partnership one, humans will not cease to exist. But it would result in a lot of unnecessary suffering, including early death for some, as well as massive biodiversity loss and possible ecosystem breakdown. On the other hand, limits to growth are providing us with a now-or-never opportunity to create purpose for ourselves by striving for human and ecological thriving. To establish a global equilibrium with and for nature in all its forms, simply because we love life more than growth.

I don't know whether humanity will make the necessary radical improvements to avoid collapse. There are many signs in both directions, a few of which I have mentioned in this book. But I do know that we can. We still have time, even though that is running out. We have the capabilities, including in technology and finance, which will be crucially helpful once we direct them to solve the right problems. Most importantly, we have the will, because striving for balance and caring for life is in our nature. We can be thrown off that natural state through force justified by stories of qualifying hierarchies, like violence, shaming, or chemical manipulation. And in a society with expansionary growth as its pursuit, we inevitably have been. But the longing to connect remains. So, if you take only one thing away from this book, let it be this: You are more and better than what you have been told.

Appendix A

Depiction of World3.

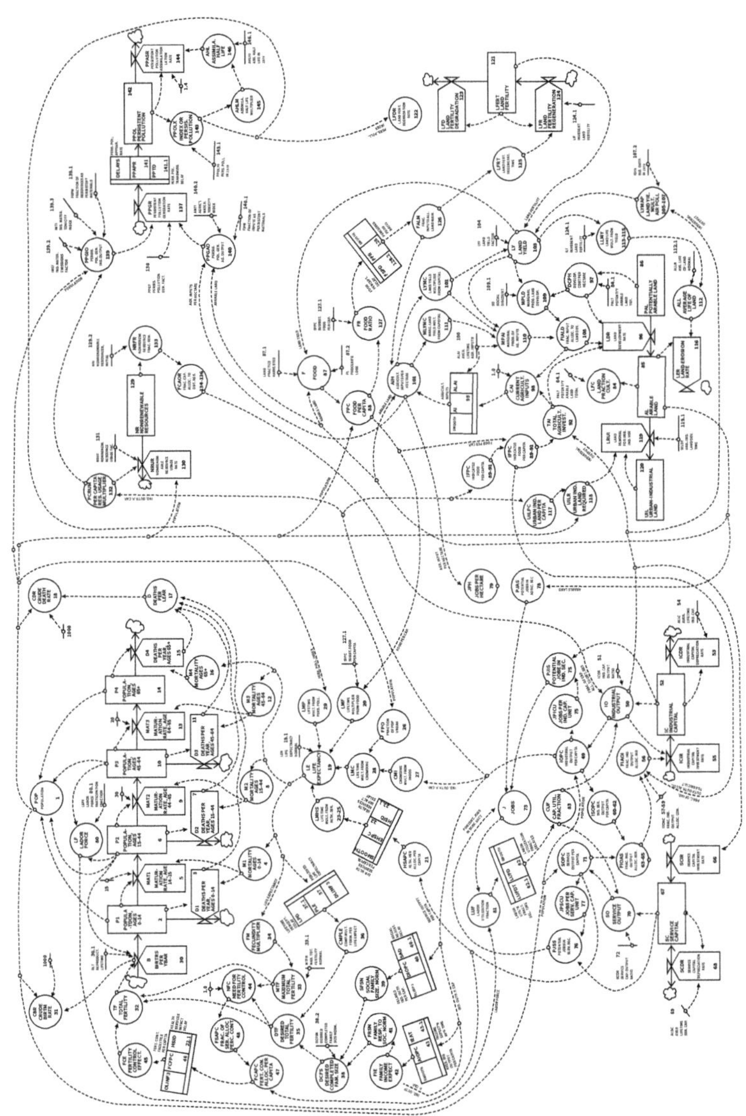

Figure A1. Depiction of the interactions in the World3 model. Source: Reprinted from Pasqualino et al. (2015).

References

African Development Bank Group (ADBG). 2020. The "High 5s": A Strategic Vision and Results That Are Transforming AFRICA. Available online: https://www.afdb.org/en/news-and-events/high-5s-strategic-vision-and-results-are-transforming-africa-35851#:~{}:text=The%20African%20Development%20Bank%E2%80%99s%20High,for%20the%20people%20of%20Africa (accessed on 17 December 2021).

Ahmed, Nabil, Anna Marriott, Nafkote Dabi, Megan Lowthers, Max Lawson, and Leah Mugehera. 2022. Inequality kills: The unparalleled action needed to combat unprecedented inequality in the wake of COVID-19. *Oxfam International.* Available online: https://www.oxfam.org/en/research/inequality-kills (accessed on 10 May 2022).

Aked, Jody, Nic Marks, Corrina Cordon, and Sam Thompson. 2008. Five Ways to Wellbeing: Communicating the Evidence. Available online: https://neweconomics.org/2008/10/five-ways-to-wellbeing (accessed on 2 May 2022).

Allaire, Maura, Haowei Wu, and Upmanu Lall. 2018. National trends in drinking water quality violations. *Proceedings of the National Academy of Science* 115: 2078–83. [CrossRef] [PubMed]

Allen, Cameron, Graciela Metternicht, Thomas Wiedmann, and Matteo Pedercini. 2021. Modelling national transformations to achieve the SDGs within planetary boundaries in small island developing states. *Global Sustainability* 4: E15. [CrossRef]

Alston, Philip. 2017. Statement on visit to the USA. Available online: http://www.ohchr.org/EN/NewsEvents/Pages/DisplayNews.aspx?NewsID=22533&LangID=E (accessed on 11 November 2021).

Amini, M. Hadi, Kianoosh G. Boroojeni, S. S. Iyengar, Panos M. Pardalos, Frede Blaabjerg, and Asad M. Madni. 2018. *Sustainable Interdependent Networks: From Theory to Application.* New York: Springer.

Arekrans, Johan, Sofia Ritzén, and Rafael Laurenti. 2022. The role of radical innovation in circular strategy deployment. *Business Strategy and the Environment*, 1–21. [CrossRef]

Armstrong McKay, David, Arie Staal, Jesse F. Abrams, Ricarda Winkelmann, Boris Sakschewski, Sina Loriani, Ingo Fetzer, Sarah E. Cornell, Johan Rockström, and Timothy M. Lenton. 2022. Exceeding 1.5 °C global warming could trigger multiple climate tipping points. *Science* 377. [CrossRef]

Aspiration Bank. 2022. Who We Are. Available online: https://www.aspiration.com/who-we-are (accessed on 11 March 2022).

Associated Press, and David Martosko. 2019. 'JOBS! JOBS! JOBS!' Trump Gloats as March Labor Market Beats Expectations with 196,000 New Jobs and Unemployment Holds at Lowest Level Since 2008. *Daily Mail.* April 5. Available online: https://www.dailymail.co.uk/news/article-6891013/Stocks-tick-higher-US-jobs-report-hits-sweet-spot.html (accessed on 3 November 2021).

Azoulay, David, Priscilla Villa, Yvette Arellano, Miriam Gordon, Doun Moon, Kathryn Miller, and Kristen Thompson. 2019. Plastic & Health: The Hidden Costs of a Plastic Planet. Available online: https://www.ciel.org/wp-content/uploads/2019/02/Plastic-and-Health-The-Hidden-Costs-of-a-Plastic-Planet-February-2019.pdf (accessed on 21 November 2021).

Baek, Jungho, and Guankerwon Gweisah. 2013. Does income inequality harm the environment?: Empirical evidence from the United States. *Energy Policy* 62: 1434–37. [CrossRef]

Bailey, Ronald. 1989. Dr. Doom. *Forbes*, October 16, 5.

Banerjee, Abhijit V., and Esther Duflo. 2019. *Good Economics for Hard Times*. New York: PublicAffairs.

Barabási, Albert-László. 2003. *Linked: How Everything Is Connected to Everything Else and What It Means for Business, Science, and Everyday Life*. New York: Basic Books.

Bardach, Ann Louise. 2014. Lifestyles of the rich and parched: How the Golden State's 1 percenters are avoiding the drought. *Politico*. August 24. Available online: https://www.politico.com/magazine/story/2014/08/california-drought-lifestyles-of-the-rich-and-parched-110305#.U_z-1Usrles (accessed on 14 October 2021).

Bardi, Ugo. 2011. Cassandra's Curse: How "The Limits to Growth" Was Demonized. Available online: https://www.resilience.org/stories/2011-09-15/cassandras-curse-how-limits-growth-was-demonized/ (accessed on 11 October 2021).

Bardi, Ugo. 2014. *Extracted: How the Quest for Mineral Wealth Is Plundering the Planet*. White River Junction: Chelsea Green Publishing Co.

Barlas, Yaman. 1996. Formal aspects of model validity and validation in system dynamics. *System Dynamics Review* 12: 183–210. [CrossRef]

Barrabi, Thomas. 2022. JPMorgan loosens return to office rules for some workers after pushback: Report. *New York Post*. April 22. Available online: https://nypost.com/2022/04/27/jpmorgan-loosens-return-to-office-rules-after-pushback/ (accessed on 13 May 2022).

Barro, Robert J., and Jong Wha Lee. 2021. BarroLeeDataset. Available online: https://barrolee.github.io/BarroLeeDataSet/BLv3.html (accessed on 2 October 2021).

Barton, Dominic, James Manyika, Timothy Koller, Robert Palter, Jonathan Godsall, and Joshua Zoffer. 2017. The Economic Impact of Short-Termism. *McKinsey Global Institute*. Available online: https://www.mckinsey.com/~{}/media/mckinsey/featured%20insights/long%20term%20capitalism/where%20companies%20with%20a%20long%20term%20view%20outperform%20their%20peers/measuring-the-economic-impact-of-short-termism.pdf (accessed on 22 October 2021).

Bendix, Aria. 2022. 'Forever chemicals' stay in the air and water permanently. But scientists have found a new way to destroy them. *NBCNews*. August 18. Available online: https://www.nbcnews.com/health/health-news/new-way-destroy-pfas-forever-chemicals-rcna43528 (accessed on 27 August 2022).

Benevolent Capitalism. 2022. The Benevolent Way to Create a Business. Available online: http://accessbenevolentcapitalism.com/ (accessed on 15 June 2022).

Bergh, Andreas, Therese Nilsson, and Daniel Waldenström. 2016. *Sick of Inequality? An Introduction to the Relationship between Inequality and Health*. Cheltenham: Edward Elgar Publishing.

Berkhout, Esmé, Nick Galasso, Max Lawson, Pablo Andrés Rivero Morales, Anjela Taneja, and Diego Alejo Vázquez Pimentel. 2021. *The Inequality Virus*. Oxford: Oxfam GB. Available online: https://www.oxfamamerica.org/explore/research-publications/inequality-virus/ (accessed on 2 February 2022).

Bernays, Edward L., and Howard Walden Cutler. 1955. *The Engineering of Consent*. Norman: University of Oklahoma Press.

Bessone, Pedro, Gautam Rao, Frank Schilbach, Heather Schofield, and Mattie Toma. 2021. *The Economic Consequences of Increasing Sleep Among the Urban Poor*; Cambridge: National Bureau of Economic Research. Available online: https://doi.org/10.3386/w26746 (accessed on 13 February 2022).

Beyond GDP. 2022. What Is the 'Beyond GDP' Initiative. Available online: https://ec.europa.eu/environment/beyond_gdp/index_en.html (accessed on 5 June 2022).

Biomimicry Institute (BI). 2022a. Biomimicry Is a Practice That Learns from and Mimics the Strategies Found in Nature to Solve Human Design Challenges—And Find Hope. Available online: https://biomimicry.org/what-is-biomimicry/?gclid=CjwKCAiA3L6PBhBvEiwAINlJ9KpcetxMosSLv5gPWTUYdDga3h5h67DQ5BQBQV-81ZTcdsBMIqlOIxoCmswQAvD_BwE (accessed on 2 January 2022).

Biomimicry Institute (BI). 2022b. How Fungi Can Clean Up Pollution. Available online: https://asknature.org/strategy/the-fungi-that-clean-up-pollution/ (accessed on 2 January 2022).

Blackburn, William R. 2015. *The Sustainability Handbook: The Complete Management Guide to Achieving Social, Economic, and Environmental Responsibility*, 2nd ed. Saint Paul: West Academic Publishing.

Bloomberg. 2022. Global Environmental, Social & Governance—ESG Data. Available online: https://www.bloomberg.com/professional/dataset/global-environmental-social-governance-data/ (accessed on 12 January 2022).

Board of Governors of the Federal Reserve System. 2019. *Report on the Economic Well-Being of U.S. Households in 2018–May 2019*. Available online: https://www.federalreserve.gov/publications/2019-economic-well-being-of-us-households-in-2018-dealing-with-unexpected-expenses.htm (accessed on 2 February 2022).

Bond, Shannon, and Raquel Maria Dillon. 2022. Twitter may have hired a Chinese spy and four other takeaways from the Senate hearing. *NPR*. September 13. Available online: https://www.npr.org/2022/09/13/1122671582/twitter-whistleblower-mudge-senate-hearing (accessed on 13 September 2022).

Bowles, Samuel, and Herbert Gintis. 2002. Homo reciprocans. *Nature* 415: 125–27. [CrossRef]

Branderhorst, Gaya. 2020. Update to Limits to Growth: Comparing the World3 Model with Empirical Data. Master's Thesis, Harvard Extension School, Cambridge, MA, USA. Available online: https://dash.harvard.edu/handle/1/37364868 (accessed on 29 October 2021).

Brathwaite, Joy, Stephen Horst, and Joseph Iacobucci. 2010. Maximizing efficiency in the transition to a coal-based economy. *Energy Policy* 38: 6084–91. [CrossRef]

Bricco, Jana, and TengTeng Xu. 2019. *Interconnectedness and Contagion Analysis: A Practical Framework.* IMF Working Paper. WP/19/220. Available online: https://www.imf.org/en/Publications/WP/Issues/2019/10/11/Interconnectedness-and-Contagion-Analysis-A-Practical-Framework-48717 (accessed on 21 January 2022).

British Petroleum (BP). 2021. Statistical Review of World Energy. Available online: https://www.bp.com/en/global/corporate/energy-economics/statistical-review-of-world-energy.html (accessed on 21 October 2021).

Brown, Brené. 2012. *The Power of Vulnerability: Teachings of Authenticity, Connection, and Courage.* Louisville: Sounds True.

Brynjolfsson, Erik, and Andrew McAfee. 2014. *The Second Machine Age: Work, Progress, and Prosperity in a Time of Brilliant Technologies.* New York: Norton & Company.

BuildRemote. 2022. *JPMorgan Chase's Return to Office Policy & Timeline.* Available online: https://buildremote.co/return-to-office/jpmorgan-chase/ (accessed on 20 August 2022).

Buurtzorg. 2022. About Us. Available online: https://www.buurtzorg.com/about-us/ (accessed on 20 January 2022).

Camus, Albert. 1991. *The Myth of Sisyphus and Other Essays.* New York: Vintage Books.

Capital Institute. 2022. *Regenerative Economics.* Available online: https://capitalinstitute.org/regenerative-capitalism/ (accessed on 5 June 2022).

Carlson, Benjamin. 2010. Quote of the Day: Google CEO Compares Data Across Millennia. *The Atlantic.* Available online: https://www.theatlantic.com/technology/archive/2010/07/quote-of-the-day-google-ceo-compares-data-across-millennia/344989/ (accessed on 21 December 2021).

Carter, Irl. 2017. *Human Behavior in the Social Environment: A Social Systems Approach.* Oxfordshire: Routledge.

Carville, Olivia. 2018. The Super Rich of Silicon Valley Have a Doomsday Escape Plan. *Bloomberg.* Available online: https://www.bloomberg.com/features/2018-rich-new-zealand-doomsday-preppers/ (accessed on 21 December 2021).

Casselman, Ben, and Jim Tankersley. 2019. Democrats want to tax the wealthy: Many voters agree. *New York Times.* February 19. Available online: https://www.nytimes.com/2019/02/19/business/economy/wealth-tax-elizabeth-warren.html (accessed on 23 December 2021).

Castro, Rodrigo. 2012. Arguments on the imminence of global collapse are premature when based on simulation models. *GAIA—Ecological Perspectives on Science and Society* 21: 271–73. [CrossRef]

Centers for Disease Control and Prevention (CDC). 2021a. Preventing Adverse Childhood Experiences. Available online: https://www.cdc.gov/violenceprevention/aces/fastfact.html?CDC_AA_refVal=https%3A%2F%2Fwww.cdc.gov%2Fviolenceprevention%2Facestudy%2Ffastfact.html (accessed on 21 December 2021).

Centers for Disease Control and Prevention (CDC). 2021b. Racism Is a Serious Threat to the Public's Health. Available online: https://www.cdc.gov/healthequity/racism-disparities/index.html (accessed on 21 December 2021).

Chan, Szu Ping. 2019. 'Why economists get things wrong'. *BBC News*. November 19. Available online: https://www.bbc.co.uk/news/business-50310815.amp (accessed on 1 December 2021).

Chancel, Lucas, Thomas Piketty, Emmanuel Saez, and Gabriel Zucman. 2021. *World Inequality Report 2022*. World Inequality Lab: Available online: https://wir2022.wid.world/www-site/uploads/2021/12/WorldInequalityReport2022_Full_Report.pdf (accessed on 21 January 2022).

Chang, Ha-Joon. 2011. *23 Things They Don't Tell You about Capitalism*. New York: Bloomsbury Publishing.

Chang, Ha-Joon. 2015. *Economics: The User's Guide*. New York: Bloomsbury Publishing.

Chaudhary, Archana. 2022. India's new ESG Rules to Address Corporate Green Washing. *Bloomberg*, September 2. Available online: https://www.bloomberg.com/news/articles/2022-09-02/india-s-new-esg-rules-to-address-corporate-greenwashing?leadSource=uverify%20wall (accessed on 18 September 2022).

Chell, Conor, Samina Ullah, and Laura Roberts. 2022. *It's Official: Mandatory ESG Disclosure Is Coming to Canada*. Available online: https://www.mltaikins.com/esg/its-official-mandatory-esg-disclosure-is-coming-to-canada/ (accessed on 25 January 2022).

Chen, David, and Walter Stroup. 1993. General System Theory: Toward a Conceptual Framework for Science and Technology Education for All. *Journal of Science Education and Technology* 2: 447–59. [CrossRef]

Chichakly, Karim. 2009. Limits to Growth. Available online: https://blog.iseesystems.com/stella-ithink/limits-to-growth/ (accessed on 21 September 2021).

Chow, Denise. 2021. Triple Jeopardy: Children Face Dark Future of Climate Disasters. *NBCNews*. September 27. Available online: https://www.nbcnews.com/science/environment/triple-jeopardy-children-face-dark-future-climate-disasters-rcna2304 (accessed on 21 January 2022).

Cingano, Federico. 2014. *Trends in Income Inequality and Its Impact on Economic Growth*. OECD Social, Employment and Migration Working Papers, No. 163. Paris: OECD Publishing. [CrossRef]

Clifton, Jon. 2022. The Global Rise of Unhappiness. *Gallup*, September 15. Available online: https://news.gallup.com/opinion/gallup/401216/global-rise-unhappiness.aspx (accessed on 16 September 2022).

Club of Rome (CoR). 2022. Earth4All Project. Available online: https://www.earth4all.life/ (accessed on 27 January 2022).

Cohen, Patricia. 2015. One Company's New Minimum Wage: $70,000 a Year. *New York Times*. April 13. Available online: https://www.nytimes.com/2015/04/14/business/owner-of-gravity-payments-a-credit-card-processor-is-setting-a-new-minimum-wage-70000-a-year.html (accessed on 2 January 2022).

Cole, H. S. D., Christopher Freeman, Marie Jahoda, and K. L. R. Pavitt. 1973. *Models of Doom: A Critique of the Limits to Growth*. Bloomington: Universe Publishing.

Collste, David. 2020. New Economic Paradigms. In *Reference Module in Earth Systems and Environmental Sciences*. Edited by Scott Elias. Amsterdam: Elsevier, pp. 296–302. [CrossRef]

Coogan, Patricia, Karin Schon, Shanshan Li, Yvette Cozier, Traci Bethea, and Lynn Rosenberg. 2020. Experiences of racism and subjective cognitive function in African American women. *Alzheimer's Dement.* 12: e12067. [CrossRef]

Da Silva, Chantal. 2022. Apple workers launch petition over company's reported return-to-office plan. *NBCNews*. August 22. Available online: https://www.nbcnews.com/news/us-news/apple-workers-launch-petition-companys-reported-return-office-plan-rcna44156 (accessed on 22 August 2022).

Dabla-Norris, Era, Kalpana Kochhar, Nujin Suphaphiphat, Franto Ricka, and Evridiki Tsounta. 2015. *Causes and Consequences of Income Inequality: A Global Perspective*. Staff Discussion Notes No. 2015/013. Available online: https://www.imf.org/en/Publications/Staff-Discussion-Notes/Issues/2016/12/31/Causes-and-Consequences-of-Income-Inequality-A-Global-Perspective-42986 (accessed on 14 October 2021).

de Jongh, D. C. J. 1978. Structural parameter sensitivity of the Limits to Growth world model. *Applied Mathematical Modelling* 2: 77–80. [CrossRef]

De Nederlandsche Bank (DNB). 2022. Mission and Tasks. Available online: https://www.dnb.nl/en/about-us/mission-and-tasks/ (accessed on 11 January 2022).

de Vries, Robert, Samuel Gosling, and Jeff Potter. 2011. Income inequality and personality: Are less equal U.S. states less agreeable? *Social Science & Medicine* 72: 1978–85. [CrossRef]

Deutsche Bank. 2020. Biodiversity Loss: How Many Extinctions Add Up to Economic Collapse? Available online: https://deutschewealth.com/en/articles/biodiversity-loss-podcast.html (accessed on 19 February 2022).

Deutsche Bank. 2022. Biodiversity: The New Playing Field for ESG Assessment. Available online: https://deutschewealth.com/en/our_perspective/cio-specials/biodiversity-new-playing-field-esg-assessment.html (accessed on 19 February 2022).

Dfarhud, Dariush, Maryam Malmir, and Mohammad Khanahmadi. 2014. Happiness & Health: The Biological Factors- Systematic Review Article. *Iranian Journal of Public Health* 43: 1468–77.

Diamond, Jared. 2011. *Collapse: How Societies Choose to Fail or Succeed*. London: Penguin Books.

Diamond, Jared. 2019. *Upheaval: Turning Points for Nations in Crisis*. Boston: Little, Brown and Company.

DiAngelo, Robin. 2018. *White Fragility: Why It's So Hard for White People to Talk About Racism*. Boston: Beacon Press.

Dixson-Declève, Sandrine. 2021. 5 Keys to Shifting to Sustainable Growth—And the Cost of Inaction. Available online: https://www.ted.com/talks/sandrine_dixson_decleve_5_keys_to_shifting_to_sustainable_growth_and_the_cost_of_inaction#t-180270s (accessed on 1 February 2022).

Dixson-Declève, Sandrine, Per Espen Stoknes, Owen Gaffney, and Jayati Ghosh. 2022. *Earth for All: A Plan for Global Wellbeing on a Healthy Planet*. A Report to the Club of Rome. Gabriola: New Society.

Dlugokencky, Ed, and Pieter Tans. 2021. Globally Averaged Marine Surface Annual Mean Data. [Data File]. Available online: https://gml.noaa.gov/ccgg/trends/gl_data.html (accessed on 1 November 2021).

Dodson, Robin E., Julia O. Udesky, Meryl D. Colton, Martha McCauley, David E. Camann, Alice Y. Yau, Gary Adamkiewicz, and Ruthann A. Rudel. 2017. Chemical exposures in recently renovated low-income housing: Influence of building materials and occupant activities. *Environmental International* 109: 114–27. [CrossRef]

Doughnut Economics Action Lab (DEAL). 2022. Available online: https://doughnuteconomics.org/ (accessed on 5 February 2022).

Doughnut Economics Action Lab (DEAL), Biomimicry 3.8, Circle Economy, and C40. 2020. The Amsterdam City Doughnut: A Tool for Transformative Action. Available online: https://www.kateraworth.com/wp/wp-content/uploads/2020/04/20200406-AMS-portrait-EN-Single-page-web-420x210mm.pdf (accessed on 5 February 2022).

Dorling, Danny. 2017. *The equality effect: Improving life for everyone*. Oxford: New Internationalist Publications Ltd.

Dry Bar Comedy. 2018. Math is Make Believe. Paul Morrissey. Video. October 12. Available online: https://www.youtube.com/watch?v=2u3_enoT4xM (accessed on 1 September 2021).

Duffy, Clare. 2021. Facebook Whistleblower Revealed on '60 Minutes,' Says the Company Prioritized Profit over Public Good. *CNN Business*. October 4. Available online: https://www.cnn.com/2021/10/03/tech/facebook-whistleblower-60-minutes/index.html (accessed on 1 November 2021).

Duncan, Greg J., and Richard J. Murnane. 2014. *Restoring Opportunity: The Crisis of Inequality and the Challenge for American Education*. Cambridge: Harvard Education Press.

Dur, Robert, and Max van Lent. 2018. *Socially Useless Jobs*. Tinbergen Institute Discussion Paper 18-034/VII. Available online: https://papers.ssrn.com/sol3/papers.cfm?abstract_id=3162569 (accessed on 1 November 2021).

Eavis, Peter, and Keith Collins. 2018. The Banks Changed. Except for All the Ways They're the Same. *New York Times*. September 12. Available online: https://www.nytimes.com/interactive/2018/09/12/business/big-investment-banks-dodd-frank.html (accessed on 29 November 2021).

Eckstein, David, Vera Künzel, and Laura Schäfer. 2021. *Global Climate Risk Index 2021*. Bonn: Germanwatch e.V. Available online: https://reliefweb.int/sites/reliefweb.int/files/resources/Global%20Climate%20Risk%20Index%202021_1_0.pdf (accessed on 21 January 2022).

Edelman. 2020. 2020 Edelman Trust Barometer. Available online: https://www.edelman.com/trust/2020-trust-barometer (accessed on 12 November 2021).

Edmondson, Amy C. 2018. *The Fearless Organization: Creating Psychological Safety in the Workplace for Learning, Innovation, and Growth*. Hoboken: Wiley.

Eide, Eric R., Mark H. Showalter, and Dan D. Goldhaber. 2010. The relation between children's health and academic achievement. *Children and Youth Service Review* 32: 231–38. [CrossRef]

Eisenstein, Charles. 2018. *Climate: A New Story*. Berkeley: North Atlantic Books.

Eisler, Riane. 1988. *The Chalice and the Blade: Our History, Our Future*. New York: HarperCollins.

Eisler, Riane. 2008. *The Real Wealth of Nations: Creating A Caring Economics*. Oakland: Berrett-Koehler Publishers.

Ellison, George T. H. 1999. Income inequality, social trust, and self-reported health status in high-income countries. *Annals of the New York Academy of Sciences* 896: 325–28. [CrossRef]

Epstein, Joshua M., and Robert L. Axtell. 1996. *Growing Artificial Societies: Social Science from the Bottom Up (Complex Adaptive Systems)*. Cambridge: MIT Press.

Etheridge, David M., L. P. Steele, R. Li Langenfelds, Roger J. Francey, J.-M. Barnola, and V. I. Morgan. 1996. Natural and anthropogenic changes in atmospheric CO_2 over the last 1000 years from air in Antarctic ice and firn. *Journal of Geophysical Research* 101: 4115–28. [CrossRef]

Environmental Working Group (EWG). 2005. Body Burden: The Pollution in Newborns. Available online: https://www.ewg.org/research/body-burden-pollution-newborns (accessed on 4 January 2022).

Environmental Working Group (EWG). 2022. The 'Forever Chemicals' in 99% of Americans. Available online: https://www.ewg.org/pfaschemicals/what-are-forever-chemicals.html (accessed on 4 January 2022).

European Commission (EC). 2019. A European Green Deal. Available online: https://ec.europa.eu/info/strategy/priorities-2019-2024/european-green-deal_en (accessed on 7 July 2022).

European Commission (EC). 2022a. Corporate Sustainability Reporting. Available online: https://ec.europa.eu/info/business-economy-euro/company-reporting-and-auditing/company-reporting/corporate-sustainability-reporting_en (accessed on 7 February 2022).

European Commission (EC). 2022b. Quality of Life Indicators—Overall Experience of Life. Available online: https://ec.europa.eu/eurostat/statistics-explained/index.php?title=Quality_of_life_indicators_-_overall_experience_of_life (accessed on 1 February 2022).

Fajnzylber, Pablo, Daniel Lederman, and Norman Loayza. 2002. Inequality and violent crime. *The Journal of Law and Economics* 45: 1–39. [CrossRef]

FAO, IFAD, UNICEF, WFP, and WHO. 2021. *The State of Food Security and Nutrition in the World 2021. Transforming Food Systems for Food Security, Improved Nutrition and Affordable Healthy Diets for All*. Rome: FAO. Available online: https://www.fao.org/documents/card/en/c/cb4474en (accessed on 7 November 2021).

FAOSTAT. 2021a. Food Balances (2010-)—Metadata. Available online: http://www.fao.org/faostat/en/#data/FBS/metadata (accessed on 7 November 2021).

FAOSTAT. 2021b. Food Balances (-2013, Old Methodology and Population)—Download Data. Available online: https://www.fao.org/faostat/en/#data/FBSH (accessed on 7 November 2021).

FAOSTAT. 2021c. Food Balances (2010-)—Download Data. Available online: https://www.fao.org/faostat/en/#data/FBS (accessed on 7 November 2021).

Farinosi, Fabio, Carlo Giupponi, Arnaud Reynaud, Guido Ceccherini, César Carmona-Moreno, Ad P. J. De Roo, D. Gonzalez Sanchez, and Giovanni Bidoglio. 2018. An innovative approach to the assessment of hydro-political risk: A spatially explicit, data driven indicator of hydro-political issues. *Global Environmental Change* 52: 286–313. [CrossRef] [PubMed]

Farrow, Aidan, Kathryn A. Miller, and Lauri Myllyvirta. 2020. *Toxic Air: The Price of Fossil Fuels.* Seoul: Greenpeace Southeast Asia. Available online: https://storage.googleapis.com/planet4-southeastasia-stateless/2020/02/21b480fa-toxic-air-report-110220.pdf (accessed on 7 December 2021).

Farver, Suzanne. 2019. *Pathways to Success: Case Studies for Mainstreaming Corporate Sustainability.* Plantation: J. Ross Publishing.

Faucon, Benoît. 2013. Energy: Oil Companies Go Deep—Despite Predictions of Retrenchment, Offshore Drilling Has Taken Off. *The Wall Street Journal Asia.* November 11, p. 10. Available online: https://www.wsj.com/articles/SB10001424052702303442004579123560225082786 (accessed on 5 January 2022).

Federal Reserve Board. 2021. Second Conference on the Interconnectedness of Financial Systems. Available online: https://www.federalreserve.gov/conferences/conference-on-the-interconnectedness-of-financial-systems-202112.htm (accessed on 7 January 2022).

Financial Stability Board. 2021. 2021 List of Global Systemically Important Banks (G-SIBs). Available online: https://www.fsb.org/2021/11/2021-list-of-global-systemically-important-banks-g-sibs/ (accessed on 29 December 2021).

Ford, David N., and John D. Sterman. 1998. Expert knowledge elicitation to improve formal and mental models. *System Dynamics Review* 14: 309–40. Available online: https://citeseerx.ist.psu.edu/viewdoc/download?doi=10.1.1.559.9110&rep=rep1&type=pdf (accessed on 7 October 2021). [CrossRef]

Foroohar, Rana. 2016. *Makers and Takers: The Rise of Finance and the Fall of American Business.* New York: Penguin Random House.

Forrester, Jay W. 1971. *World Dynamics.* Cambridge: Wright-Allen Press.

Forrester, Jay W. 1975. *Collected Papers.* Waltham: Pegasus Communications.

Forrester, Jay W., Gilbert W. Low, and Nathaniel J. Mass. 1974. The debate on "World Dynamics": A response to Nordhaus. *Policy Sciences* 5: 169–90. [CrossRef]

Friedman, Benjamin M. 2017. Work and consumption in an era of unbalanced technological advance. *Journal of Evolutionary Economics* 27: 221–37. [CrossRef]

Friedman, Milton. 1970. A Friedman Doctrine—The Social Responsibility of Business Is to Increase Its Profits. *The New York Times.* September 13. Available online: https://www.nytimes.com/1970/09/13/archives/a-friedman-doctrine-the-social-responsibility-of-business-is-to.html (accessed on 7 January 2022).

Fuller, Richard, Philip J. Landrigan, Kalpana Balakrishnan, Glynda Bathan, Stephan Bose-O'Reilly, Michael Brauer, Jack Caravanos, Tom Chiles, Aaron Cohen, Lilian Corra, and et al. 2022. Pollution and health: A progress update. *The Lancet* 6: 535–47. [CrossRef]

Gaffney, Owen. 2022. We're hiring a head of communications. *LinkedIn.* August. Available online: https://www.linkedin.com/feed/update/urn:li:activity:6966302156647772160/ (accessed on 25 August 2022).

Gaffney, Owen, Zoe Tcholak-Antitch, Sophia Boehm, Stephan Barthel, Thomas Hahn, Lisa Jacobson, Kelly Levin, Diana Liverman, Per Espen Stoknes, Sophie Thompson, and et al. 2021. Global Commons Survey: Attitudes to planetary stewardship. Available online: https://globalcommonsalliance.org/wp-content/uploads/2021/08/Global-Commons-G20-Survey-full-report.pdf (accessed on 30 December 2021).

Gallup. 2021. State of the Global Workplace: 2022 Report. Available online: https://www.gallup.com/workplace/349484/state-of-the-global-workplace.aspx (accessed on 30 December 2021).

Gartenberg, Claudine, and George Serafeim. 2019. 181 Top CEOs Have Realized Companies Need a Purpose Beyond Profit. *Harvard Business Review.* August 20. Available online: https://hbr.org/2019/08/181-top-ceos-have-realized-companies-need-a-purpose-beyond-profit (accessed on 28 December 2021).

Gaspar, Vitor, and Mercedes Garcia-Escribano. 2017. Inequality: Fiscal Policy Can Make the Difference. Available online: https://blogs.imf.org/2017/10/11/inequality-fiscal-policy-can-make-the-difference/ (accessed on 29 December 2021).

Gatti, Luciana V., Luana S. Basso, John B. Miller, Manuel Gloor, Lucas Gatti Domingues, Henrique L. G. Cassol, Graciela Tejada, Luiz E. O. C. Aragão, Carlos Nobre, Wouter Peters, and et al. 2021. Amazonia as a carbon source linked to deforestation and climate change. *Nature* 595: 388–93. [CrossRef] [PubMed]

Geyer, Roland, Jenna R. Jambeck, and Kara Lavender Law. 2017. Production, use, and fate of all plastics ever made. *Science Advances* 3: 7. [CrossRef] [PubMed]

Gigerenzer, Gerd, and Henry Brighton. 2009. Homo heuristicus: Why biased minds make better inferences. *Cognitive Science Society* 1: 107–43. [CrossRef] [PubMed]

Gilbert, Natasha. 2022. Climate Change Will Force New Animal Encounters—And Boost Viral Outbreaks. *Nature.* April 28. Available online: https://www.nature.com/articles/d41586-022-01198-w (accessed on 30 January 2022).

Global Carbon Project. 2021. Carbon BUDGET and trends 2021. Available online: https://essd.copernicus.org/preprints/essd-2021-386/ (accessed on 13 January 2022).

Global Footprint Network (GFN). 2021a. Ecological Footprint. Available online: https://www.footprintnetwork.org/our-work/ecological-footprint/ (accessed on 2 December 2021).

Global Footprint Network (GFN). 2021b. FAQs. Available online: https://www. footprintnetwork.org/faq/ (accessed on 2 December 2021).

Goldstein, Jacob. 2020. *Money: The True Story of a Made-Up Thing*. Paris: Hachette Books.

Golüke, Ulrich. 2018. *Generous Respect: The Next Story of Humanity*. Norderstedt: Books on Demand.

Goodkind, Andrew L., Christopher W. Tessum, Jay S. Coggins, Jason D. Hill, and Julian D. Marshall. 2019. Fine-scale damage estimates of particulate matter air pollution reveal opportunities for location-specific mitigation of emissions. *Proceedings of the National Academy of Sciences of the United States of America* 116: 8775–80. [CrossRef] [PubMed]

Googins, Bradley K., Philip H. Mirvis, and Steven A. Rochlin. 2007. *Beyond Good Company: Next Generation Corporate Citizenship*. New York: Palgrave Macmilian.

Gould, Eric D., and Alexander Hijzen. 2016. Growing Apart, Losing Trust? The Impact of Inequality on Social Capital. Available online: https://www.imf.org/external/pubs/ft/ wp/2016/wp16176.pdf (accessed on 4 October 2021).

Government of Canada. 2021. Measuring What Matters: Toward a Quality of Life Strategy for Canada. Available online: https://www.canada.ca/en/department-finance/services/ publications/measuring-what-matters-toward-quality-life-strategy-canada.html (accessed on 16 January 2022).

Grabowska, Sandra, Sebastian Saniuk, and Bożena Gajdzik. 2022. Industry 5.0: Improving humanization and sustainability of Industry 4.0. *Scientometrics* 127: 3117–44. [CrossRef] [PubMed]

Graeber, David. 2018. *Bullshit Jobs: A Theory*. New York: Simon & Schuster.

Graeber, David, and David Wengrow. 2021. *The Dawn of Everything: A New History of Humanity*. New York: Farrar, Straus and Giroux.

Graedel, T. E., E. M. Harper, Nedal T. Nassar, and Barbara K. Reck. 2015. On the materials basis of modern society. *Proceedings of the National Academy of Sciences* 112: 6295–300. [CrossRef] [PubMed]

Greenhouse, Steven. 2022. Is organized labor making a comeback? *The Atlantic*. April 4. Available online: https://www.theatlantic.com/ideas/archive/2022/04/how-build-union-victory-amazon-staten-island/629464/ (accessed on 29 May 2022).

Groves, David G., James Syme, Edmundo Molina-Perez, Carlos Calvo Hernandez, Luis F. Víctor-Gallardo, Guido Godinez-Zamora, Jairo Quirós-Tortós, Felipe De León Denegri, Andrea Meza Murillo, Valentina Saavedra Gómez, and et al. 2020. *The Benefits and Costs of Decarbonizing Costa Rica's Economy: Informing the Implementation of Costa Rica's National Decarbonization Plan Under Uncertainty*. Santa Monica: RAND Corporation. Available online: https://www.rand.org/pubs/research_reports/RRA633-1.html (accessed on 18 June 2022).

Haidt, Jonathan. 2013. *The Righteous Mind: Why Good People Are Divided by Politics and Religion*. New York: Pantheon.

Halden, Rolf U. 2010. Plastics and Health Risks. *Annual Review of Public Health* 31: 179–94. [CrossRef] [PubMed]

Hall, Charles A. S., and John W. Day. 2009. Revisiting the Limits to Growth after peak oil: In the 1970s a rising world population and the finite resources available to support it were hot topics. Interest faded—but it's time to take another look. *American Scientist* 97: 230–37. [CrossRef]

Hamel, Gary, and Michele Zanini. 2020. *Humanocracy: Creating Organizations as Amazing as the People Inside Them*. Brighton: Harvard Business Review Press.

Happy Planet Index. 2022. Available online: https://happyplanetindex.org/ (accessed on 1 July 2022).

Harari, Yuval Noah. 2015. *Sapiens: A Brief History of Humankind*. New York: HarperCollins Publishers.

Hardin, Garrett. 1968. The Tragedy of the Commons. *Science* 162: 1243–48. [CrossRef] [PubMed]

Hare, Brian, and Vanessa Woods. 2021. *Survival of the Friendliest: Understanding Our Origins and Rediscovering Our Common Humanity*. New York: Penguin Random House.

Harper, Sam, Corinne A. Riddell, and Nicholas B. King. 2021. Declining Life Expectancy in the United States: Missing the Trees for the Forest. *Annual Review of Public Health* 42: 381–403. [CrossRef] [PubMed]

Harter, Jim. 2022. *Percent Who Feel Employer Cares About Their Wellbeing Plummets*. Washington, DC: Gallup. Available online: https://www.gallup.com/workplace/390776/percent-feel-employer-cares-wellbeing-plummets.aspx (accessed on 9 February 2022).

Hébert-Dufresne, Laurent, Timothy M. Waring, Guillaume St-Onge, Meredith T. Niles, Laura Kati Corlew, Matthew P. Dube, Stephanie J. Miller, Nicholas J. Gotelli, and Brian J. McGill. 2022. Source-sink behavioural dynamics limit institutional evolution in a group-structured society. *Royal Society Open Science* 9: 211743. [CrossRef]

Heinze, Christoph, Thorsten Blenckner, Helena Martins, Dagmara Rusiecka, Ralf Döscher, Marion Gehlen, Nicolas Gruber, Elisabeth Holland, Øystein Hov, Fortunat Joos, and et al. 2021. The quiet crossing of ocean tipping points. *Proceedings of the National Academy of Sciences of the United States of America* 118: e2008478118. [CrossRef] [PubMed]

Helm, Dieter. 2011. Peak oil and energy policy—A critique. *Oxford Review of Economic Policy* 27: 68–91. [CrossRef]

Henckens, Matheus L. C. M., Ekko C. van Ierland, Peter P. J. Driessen, and Ernst Worrell. 2016. Mineral resources: Geological scarcity, market price trends, and future generations. *Resources Policy* 49: 102–11. [CrossRef]

Henderson, Rebecca. 2020. *Reimagining Capitalism in a World on Fire*. New York: PublicAffairs.

Hernández-Blanco, Marcello, and Robert Costanza. 2021. A new economics to achieve sustainable development goals. In *Oxford Research Encyclopedia of Environmental Science*. Oxford: Oxford University Press. [CrossRef]

Herrington, Gaya. 2020. Update to limits to growth: Comparing the world3 model with empirical data. *Journal of Industrial Ecology* 25: 614–26. [CrossRef]

Holacracy. 2022. Available online: https://www.holacracy.org/ (accessed on 22 January 2022).

Hollingsworth, Julia. 2020. How New Zealand Became an Apocalypse Escape Destination for Americans. *CNN*. July 17. Available online: https://www.cnn.com/2020/07/15/business/bunkers-new-zealand-intl-hnk/index.html (accessed on 2 November 2021).

Homan, Patricia. 2019. Structural Sexism and Health in the United States: A New Perspective on Health Inequality and the Gender System. *American Sociological Review* 84: 486–516. [CrossRef]

Intergovernmental Panel on Climate Change. 2021. Climate Change 2021: The Physical Science Basis. Contribution of Working Group I to the Sixth Assessment Report of the Intergovernmental Panel on Climate Change. Available online: https://www.ipcc.ch/report/ar6/wg1/ (accessed on 22 February 2022).

Intergovernmental Panel on Climate Change. 2022. Climate Change 2022: Impacts, Adaptation and Vulnerability. Contribution of Working Group I to the Sixth Assessment Report of the Intergovernmental Panel on Climate Change. Available online: https://www.ipcc.ch/report/ar6/wg2/ (accessed on 22 May 2022).

International Energy Agency (IEA). 2022. Global CO_2 Emissions Rebounded to Their Highest Level in History in 2021. Available online: https://www.iea.org/news/global-co2-emissions-rebounded-to-their-highest-level-in-history-in-2021 (accessed on 19 May 2022).

International Monetary Fund (IMF). 2022. *World Economic Outlook: War Sets Back the Global Recovery*; Washington, DC: IMF. Available online: https://www.imf.org/en/Publications/WEO/Issues/2022/04/19/world-economic-outlook-april-2022 (accessed on 21 May 2022).

International Renewable Energy Agency (IRENA). 2021. Renewable Power Generation Costs in 2020. Available online: https://www.irena.org/publications/2021/Jun/Renewable-Power-Costs-in-2020 (accessed on 29 January 2022).

Islam, S. Nazrul. 2015. *Inequality and Environmental Sustainability*. New York: United Nations, Department of Economic & Social Affairs. Available online: http://www.un.org/esa/desa/papers/2015/wp145_2015.pdf (accessed on 19 October 2021).

Jackson, Sarah. 2022. A Massive 4-Day Work Week Trial in the UK Began in June. Many Companies Involved say It's Been a Boon for Business and Workers Alike. *Insider*, September 21. Available online: https://www.businessinsider.com/four-day-work-week-trial-uk-halfway-mark-points-success-2022-9 (accessed on 23 September 2022).

Jackson, Tim. 2005. Live better by consuming less? Is there a 'double dividend' in sustainable consumption? *Industrial Ecology* 9: 19–36. [CrossRef]

Jackson, Tim, and Robin Weber. 2016. Limits Revisited: A Review of the Limits to Growth Debate. Available online: https://limits2growth.org.uk/revisited/ (accessed on 29 October 2021).

Jahan, Sarwat, and Ahmed Saber Mahmud. 2015. What Is Capitalism? *Finance & Development* 52. Available online: https://www.imf.org/external/pubs/ft/fandd/2015/06/basics.htm (accessed on 29 September 2021).

Jakob, Michael, and Jérôme Hilaire. 2015. Unburnable fossil-fuel reserves. *Nature* 517: 150–52. [CrossRef] [PubMed]

Jebb, Andrew T., Louis Tay, Ed Diener, and Shigehiro Oishi. 2018. Happiness, income satiation and turning points around the world. *Nature Human Behaviour* 2: 33–38. [CrossRef] [PubMed]

Jones, Trina. 2017. A different class of care: The benefits crisis of low-wage workers. *American University Law Review* 66: 691–760. Available online: http://heinonline.org/HOL/LandingPage?handle=hein.journals/aulr66&div=22&id=&page= (accessed on 30 October 2021). [PubMed]

Joshi, Ekta. 2020. A dog whistle for the billionaires – It is crafted with 470 diamonds and costs \$135,000. *Luxurylaunches*. November 23. Available online: https://luxurylaunches.com/pets/a-dog-whistle-for-the-billionaires-it-is-crafted-with-470-diamonds-and-costs-135000.php (accessed on 30 October 2021).

Kahneman, Daniel. 2011. *Thinking, Fast and Slow*. New York: Farrar, Straus and Giroux.

Kahneman, Daniel, and Amos Tversky. 1979. Prospect Theory: An Analysis of Decision under Risk. *Econometrica* 47: 263–91. [CrossRef]

Kahneman, Daniel, Olivier Sibony, and Cass R. Sunstein. 2021. *Noise: A Flaw in Human Judgment*. New York: Little, Brown Spark.

Kanter, Beth, and Aliza Sherman. 2017. *The Happy, Healthy Nonprofit: Strategies for Impact without Burnout*. Hoboken: Wiley.

Karlsson, Niklas, George Loewenstein, and Duane Seppi. 2009. The Ostrich Effect: Selective Attention to Information. *Journal of Risk and Uncertainty* 38: 95–115. [CrossRef]

Katz, Jackson. 2012. Memo to Media: Manhood, Not Guns or Mental Illness, Should Be Central in Newtown Shooting. *Huffpost*. December 18. Available online: https://www.huffpost.com/entry/men-gender-gun-violence_b_2308522 (accessed on 30 November 2021).

Kaysen, Carl. 1972. The computer that printed out W*O*L*F*. *Foreign Affairs* 50: 660–68. [CrossRef]

Keen, Steve. 2011. Debunking Macroeconomics. *Economic Analysis and Policy* 3: 147–67. [CrossRef]

Keen, Steve. 2021. *The New Economics: A Manifesto*. Cambridge: Polity.

Keister, Todd. 2010. *Bailouts and Financial Fragility*. Federal Reserve Bank of New York Staff Reports. No. 473. Available online: https://www.newyorkfed.org/medialibrary/media/research/staff_reports/sr473.pdf (accessed on 30 December 2021).

Kellner, Douglas. 2008. *Guys and Guns Amok: Domestic Terrorism and School Shootings from the Oklahoma City Bombing to the Virginia Tech Massacre*. Oxfordshire: Routledge.

Kendi, Ibram X. 2019. *How to Be an Antiracist*. New York: Penguin Random House.

Kenrick, Douglas T., Vladas Griskevicius, Jill M. Sundie, Norman P. Li, Yexin Jessica Li, and Steven L. Neuberg. 2009. Deep Rationality: The Evolutionary Economics of Decision Making. *Social Cognition* 27: 764–85. [CrossRef] [PubMed]

Ketcham, Christopher. 2018. The fallacy of endless economic growth. *Pacific Standard*. September 22. Available online: https://psmag.com/magazine/fallacy-of-endless-growth (accessed on 20 January 2022).

Keynes, John Maynard. 2010. Economic Possibilities for our Grandchildren. In *Essays in Persuasion*. New York: W.W. Norton & Co., pp. 358–332. Available online: https://link.springer.com/chapter/10.1007/978-1-349-59072-8_25 (accessed on 19 December 2021). First published 1930.

Killingsworth, Matthew A. 2021. Experienced well-being rises with income, even above $75,000 per year. *Proceedings of the National Academy of Sciences of the United States of America* 118: e2016976118. [CrossRef]

Kim, Daniel. 2021. *Introduction to Systems Thinking*. Available online: https://thesystemsthinker.com/introduction-to-systems-thinking/ (accessed on 6 October 2021).

Klein, Naomi. 2015. *This Changes Everything: Capitalism vs. The Climate*. New York: Simon & Schuster.

Kluver, Jesse, Rebecca Frazier, and Jonathan Haidt. 2014. Behavioral ethics for Homo economicus, Homo heuristicus, and Homo duplex. *Organizational Behavior and Human Decision Processes* 123: 150–58. [CrossRef]

Knight, Kyle W., and Eugene A. Rosa. 2011. The environmental efficiency of well-being: A cross-national analysis. *Social Science Research* 40: 931–49. [CrossRef]

Kose, Ayhan, Kenneth Rogoff, Eswar S. Prasad, and Shang-Jin Wei. 2003. *Effects of Financial Globalization on Developing Countries*. Washington, DC: International Monetary Fund. Available online: https://www.elibrary.imf.org/view/books/084/01998-9781589062214-en/01998-9781589062214-en-book.xml (accessed on 16 November 2021).

Kosuth, Mary, Elizabeth V. Wattenberg, Sherri A. Mason, Christopher Tyree, and Dan Morrison. 2017. *Synthetic Polymer Contamination in Global Drinking Water*. Available online: https://orbmedia.org/invisibles-final-report (accessed on 16 December 2021).

Krieger, Tim, and Daniel Meierrieks. 2016. Political capitalism: The interaction between income inequality, economic freedom, and democracy. *European Journal of Political Economy* 45: 115–32. [CrossRef]

Krippner, Greta R. 2012. *Capitalizing on Crisis: The Political Origins of the Rise of Finance*. Cambridge: Harvard University Press.

Kubiszewski, Ida. 2019. The Genuine Progress Indicator: A Measure of Net Economic Welfare. *Encyclopedia of Ecology* 327: 335–34. [CrossRef]

Kuhn, Thomas S. 1962. *The Structure of Scientific Revolutions*. Chicago: University of Chicago Press.

Kunst, Jonas R., Ronald Fischer, Jim Sidanius, and Lotte Thomsen. 2017. Hegemony preferences track societal inequality. *Proceedings of the National Academy of Sciences* 114: 5407–12. [CrossRef] [PubMed]

Laloux, Frederic. 2014. *Reinventing Organizations. A Guide to Creating Organizations Inspired by the Next Stage of Human Consciousness*. Brussels: Nelson Parker.

Lawder, David. 2019. New IMF Chief Georgieva Warns of "Synchronized Slowdown" in Global Growth. *Nasdaq*. October 8. Available online: https://www.nasdaq.com/articles/new-imf-chief-georgieva-warns-of-synchronized-slowdown-in-global-growth-2019-10-08 (accessed on 9 February 2022).

Lehmann, Geniece M., Judy S. LaKind, Matthew H. Davis, Erin P. Hines, Satori A. Marchitti, Cecilia Alcala, and Matthew Lorber. 2018. Environmental Chemicals in Breast Milk and Formula: Exposure and Risk Assessment Implications. *Environmental Health Perspectives* 126. [CrossRef] [PubMed]

Lenton, Timothy M., Johan Rockström, Owen Gaffney, Stefan Rahmstorf, Katherine Richardson, Will Steffen, and Hans Joachim Schellnhuber. 2019. Climate tipping points—Too risky to bet against. *Nature* 575: 592–95. [CrossRef]

Lewis, Simon L., and Mark A. Maslin. 2015. Defining the Anthropocene. *Nature* 519: 171–80. [CrossRef]

Lockhart, Charlotte. 2022. UK Four-Day Week Pilot Begins with 70 Companies and 3300 Workers. *4 Day Week Global.* Available online: https://www.4dayweek.com/news-posts/uk-four-day-week-pilot-begins (accessed on 21 June 2022).

Lomborg, Bjorn, and Olivier Rubin. 2009. The dustbin of history: Limits to Growth. *Foreign Policy.* November 9. Available online: https://foreignpolicy.com/2009/11/09/the-dustbin-of-history-limits-to-growth/ (accessed on 13 October 2021).

Lovins, Hunter, Amory Lovins, and Paul Hawken. 1999. *Natural Capitalism: Creating the Next Industrial Revolution.* Oxfordshire: Routledge.

Lustig, Robert H. 2018. *The Hacking of the American Mind: The Science Behind the Corporate Takeover of Our Bodies and Brains.* New York: Penguin Random House.

Luyendijk, Joris. 2015. *Swimming with Sharks.* London: Guardian faber publishing.

Lyneis, James M. 2000. System dynamics for market forecasting and structural analysis. *System Dynamics Review* 16: 3–25. [CrossRef]

Mackey, John, and Rajendra Sisodia. 2013. *Conscious Capitalism: Liberating the Heroic Spirit of Business.* Brighton: Harvard Business Review Press.

Maddison, Angus. 2006. *The World Economy—Volume 1: A Millennial Perspective and Volume 2: Historical Statistics (Development Centre Studies).* Paris: OECD Publishing.

Magnani, Elisabetta. 2000. The environmental Kuznets Curve, environmental protection policy and income distribution. *Ecological Economics* 32: 431–43. [CrossRef]

Mani, Anandi, Sendhil Mullainathan, Eldar Shafir, and Jiaying Zhao. 2013. Poverty Impedes Cognitive Function. *Science* 341: 976–80. [CrossRef] [PubMed]

Manne, Kate. 2018. *Down Girl: The Logic of Misogyny.* New York, NY: Oxford University Press.

Manno, Jack. 2002. Commoditization: Consumption efficiency and an economy of care and connection. In *Confronting Consumption.* Edited by Thomas Princen, Michael Maniates and Ken Conca. Cambridge: The MIT Press, pp. 23–42.

Martinko, Katherine. 2021. Outerwear Company Houdini Sets a High Bar for Sustainable Production. *Treehugger,* November 5. Available online: https://www.treehugger.com/outerwear-houdini-high-bar-sustainable-production-5208562 (accessed on 17 June 2022).

Maslow, A. H. 1943. A theory of human motivation. *Psychological Review* 50: 370–96. [CrossRef]

Mason, Paul. 2017. *Postcapitalism: A Guide to Our Future.* New York: Farrar, Straus and Giroux.

Masters, Jeff. 2022. Recklessness defined: Breaking 6 of 9 planetary boundaries of safety. *Yale Climate Connections*. July 12. Available online: https://yaleclimateconnections.org/2022/07/recklessness-defined-breaking-6-of-9-planetary-boundaries-of-safety/ (accessed on 8 January 2022).

May, Robert M., Simon A. Levin, and George Sugihara. 2008. Ecology for bankers. *Nature* 451: 893–94. [CrossRef] [PubMed]

Mazzucato, Mariana. 2015. *The Entrepreneurial State: Debunking Public vs. Private Sector Myths.* New York: PublicAffairs.

Mazzucato, Mariana. 2018. *The Value of Everything: Making and Taking in the Global Economy.* New York: PublicAffairs.

Mazzucato, Mariana. 2021. *Mission Economy: A Moonshot Guide to Changing Capitalism.* New York: Harper Business.

McAfee, Andrew, and Erik Brynjolfsson. 2012. Big Data: The Management Revolution. *Harvard Business Review* 90: 60–6. [PubMed]

McCormick, Barrett L. 2000. Modernization, Democracy, and Morality: The Work of Barrington Moore, Jr. *International Journal of Politics, Culture, and Society* 13: 591–606. [CrossRef]

McCurry, Justin. 2022. 'A new way of life': The Marxist, post-capitalist, green manifesto captivating Japan. *The Guardian.* September 9. Available online: https://www.theguardian.com/world/2022/sep/09/a-new-way-of-life-the-marxist-post-capitalist-green-manifesto-captivating-japan (accessed on 13 September 2022).

McDermott, Elizabeth, Jacqui Gabb, Rachael Eastham, and Ali Hanbury. 2021. Family trouble: Heteronormativity, emotion work and queer youth mental health. *Health* 25: 177–95. [CrossRef] [PubMed]

McGinnis, J. Michael. 2016. Income, life expectancy, and community health: Underscoring the opportunity. *Journal of the American Medical Association* 315: 1709–10. [CrossRef] [PubMed]

McPhillips, Deirdre. 2022. US life expectancy lowest in decades after dropping nearly a full year in 2021. *CNN.* August 31. Available online: https://www.cnn.com/2022/08/31/health/life-expectancy-declines-2021/index.html (accessed on 31 August 2022).

McSweeney, Eoin, and Adam Pourahmadi. 2021. 2% of Elon Musk's Wealth Could Help Solve World Hunger, Says Director of UN Food Scarcity Organization. *CNN Business.* November 1. Available online: https://www.cnn.com/2021/10/26/economy/musk-world-hunger-wfp-intl/index.html (accessed on 8 January 2022).

Meadows, Dennis L. 1974. *Dynamics of Growth in a Finite World.* Cambridge: Wright-Allen Press.

Meadows, Donella H. 2012. *Thinking in Systems.* White River Junction: Chelsea Green Publishing.

Meadows, Donella H., and Dennis L. Meadows. 2007. The history and conclusions of The Limits to Growth. *System Dynamics Review* 23: 191–97. [CrossRef]

Meadows, Donella H., Dennis L. Meadows, Jørgen Randers, and William W. Behrens III. 1972. *The Limits to Growth: A Report for the Club of Rome's Project on the Predicament of Mankind.* New York: Universe Books.

Meadows, Donella H., Dennis L. Meadows, and Jørgen Randers. 1992. *Beyond the Limits: Confronting Global Collapse, Envisioning a Sustainable Future.* White River Junction: Chelsea Green Publishing Co.

Meadows, Donella H., Dennis L. Meadows, and Jørgen Randers. 2004. *The Limits to Growth: The 30-Year Update.* White River Junction: Chelsea Green Publishing Co.

Meadows, Donella H., Dennis L. Meadows, and Jørgen Randers. n.d. A Synopsis: Limits to Growth: The 30-Year Update. Available online: http://donellameadows.org/archives/a-synopsis-limits-to-growth-the-30-year-update/ (accessed on 29 September 2021).

Meinema, André. 2018. Wat heeft financiële crisis ons gekost én gebracht? *NOS.* September 13. Available online: https://nos.nl/artikel/2250245-wat-heeft-financiele-crisis-ons-gekost-en-gebracht (accessed on 12 November 2021).

Melore, Chris. 2022. Money matters: 3 in 5 Americans admit waking up in middle of night thinking about finances. *StudyFinds.* April 3. Available online: https://www.studyfinds.org/money-matters-waking-up-finances/ (accessed on 1 June 2022).

MetaSD. 2022. world3-03. Available online: https://metasd.com/tag/world3/ (accessed on 1 November 2021).

Mettler, Suzanne. 2014. *Degrees of Inequality: How the Politics of Higher Education Sabotaged the American Dream.* New York: Basic Books.

Miller, James Grier. 1978. *Living Systems.* New York: McGraw-Hill.

Mima, Mahmoud, David Greenwald, and Samuel Ohlander. 2018. Environmental Toxins and Male Fertility. *Current Urology Reports* 19: 50. [CrossRef] [PubMed]

Minsky, Hyman P. 1992. *The Financial Instability Hypothesis.* The Jerome Levy Economics Institute Working Paper No. 74. Available online: https://doi.org/10.2139/ssrn.161024 (accessed on 1 October 2021). [CrossRef]

Mishel, Lawrence, and Jori Kandra. 2021. *CEO pay has skyrocketed 1,322% since 1978.* Washington, DC: Economic Policy Institute. Available online: https://www.epi.org/publication/ceo-pay-in-2020/ (accessed on 13 February 2022).

Moles, Peter, and Nicholas Terry. 1997. *The Handbook of International Financial Terms.* Oxford: Oxford University Press.

Monbiot, George. 2011. The 1% are the very best destroyers of wealth the world has ever seen. *The Guardian.* November 7. Available online: https://www.theguardian.com/commentisfree/2011/nov/07/one-per-cent-wealth-destroyers (accessed on 19 November 2021).

Monbiot, George. 2017. *Out of the Wreckage: A New Politics for an Age of Crisis.* London: Verso Books.

Morris, Kathy. 2020. Percent of People Napping at Work [survey]. *Zippia: The Career Expert.* May 11. Available online: https://www.zippia.com/advice/work-from-home-napping-survey/ (accessed on 9 June 2022).

Movahed, Masoud. 2019. The East Asian Miracle: Where Did Adam Smith Go Wrong? *Harvard International Review*. Available online: https://hir.harvard.edu/the-east-asian-miracle-where-did-adam-smith-go-wrong/ (accessed on 2 November 2021).

Moynihan, Lydia. 2022. Three-quarters of Wall Street's junior bankers want to quit their jobs: Survey. *New York Post*. August 23. Available online: https://nypost.com/2022/08/23/three-quarters-of-wall-streets-junior-bankers-want-to-quit-survey/ (accessed on 23 August 2022).

Mueller, Bernardo. 2020. Why public policies fail: Policymaking under complexity. *Economi A* 21: 311–23. [CrossRef]

Müller, Jochem Wilfried. 2021. Education and inspirational intuition—Drivers of innovation. *Heliyon* 7: E07923. [CrossRef] [PubMed]

Murtin, Fabrice. 2013. Long-term determinants of the demographic transition, 1870–2000. *Review of Economics and Statistics* 95: 617–31. [CrossRef]

Naidu, Ravi, Bhabananda Biswas, Ian R. Willett, Julian Cribb, Brajesh Kumar Singh, C. Paul Nathanail, Frederic Coulon, Kirk T. Semple, Kevin C. Jones, Adam Barclay, and et al. 2021. Chemical pollution: A growing peril and potential catastrophic risk to humanity. *Environment International* 156: 106616. [CrossRef] [PubMed]

Nair, Chandran. 2018. *The Sustainable State: The Future of Government, Economy, and Society*. Oakland: Berrett-Koehler Publishers.

Nair, Chandran. 2022. *Dismantling Global White Privilege: Equity for a Post-Western World*. Oakland: Berrett-Koehler Publishers.

Nair, Swathi. 2021. Climate Inaction Costlier Than Net Zero Transition: Reuters Poll. *Reuters*. October 25. Available online: https://www.reuters.com/business/cop/climate-inaction-costlier-than-net-zero-transition-economists-2021-10-25/ (accessed on 9 March 2022).

Natali, Susan M., John P. Holdren, Brendan M. Rogers, Rachael Treharne, Philip B. Duffy, Rafe Pomerance, and Erin MacDonald. 2021. Permafrost carbon feedbacks threaten global climate goals. *Proceedings of the National Academy of Sciences* 118: e2100163118. [CrossRef] [PubMed]

National Research Council. 2014. *The Growth of Incarceration in the United States: Exploring Causes and Consequences*. Washington, DC: The National Academies Press.

Neal, Larry, and Rondo Cameron. 2015. *A Concise Economic History of the World: From Paleolithic Times to the Present*. Oxford: Oxford University Press.

Newman, Mark, Albert-László Barabási, and Duncan J. Watts. 2005. *The Structure and Dynamics of Networks*. Princeton: Princeton University Press.

Newport, Frank. 2018. Democrats More Positive about Socialism than Capitalism. *Gallup*. August 13. Available online: https://news.gallup.com/poll/240725/democrats-positive-socialism-capitalism.aspx (accessed on 29 December 2021).

Nizzetto, Luca, Sindre Langaas, and Martyn Futter. 2016. Pollution: Do microplastics spill on to farm soils? *Nature* 537: 488. [CrossRef] [PubMed]

National Oceanic and Atmospheric Administration (NOAA). 2021. NOAA/GML Calculation of Global Means. Available online: https://www.esrl.noaa.gov/gmd/ccgg/about/global_means.html (accessed on 9 November 2021).

Nordell, Jessica. 2021. This Is How Everyday Sexism Could Stop You from Getting That Promotion. *The New York Times*. October 14. Available online: https://www.nytimes.com/interactive/2021/10/14/opinion/gender-bias.html (accessed on 1 November 2021).

Nordhaus, William D. 1973. World dynamics: Measurement without data. *The Economic Journal* 83: 1156–83. [CrossRef]

Nordhaus, William D. 1992. Lethal model 2: Limits to Growth Revisited. *Brookings Papers on Economic Activity*. Available online: https://www.brookings.edu/bpea-articles/lethal-model-2-the-limits-to-growth-revisited/ (accessed on 9 September 2021).

Norgard, Jørgen Stig, John Peet, and Kristín Vala Ragnarsdóttir. 2010. The history of limits to growth. *Solutions* 1: 59–63.

NPR Staff. 2011. Harvard Economics Students Protest Perceived Bias. *NPR*. November 3. Available online: https://www.npr.org/2011/11/03/141969009/economics-class-protests-perceived-bias (accessed on 17 October 2021).

NPR Staff. 2019. Transcript: Greta Thunberg's speech at the U.N. *NPR*. September 23. Available online: https://www.npr.org/2019/09/23/763452863/transcript-greta-thunbergs-speech-at-the-u-n-climate-action-summit?t=1658930108451 (accessed on 9 October 2021).

O'Brien, Diana, Andy Main, Suzanne Kounkel, and Anthony R. Stephan. 2019. Purpose Is Everything: How Brands That Authentically Lead with Purpose Are Changing the Nature of Business Today. *Deloitte*. October 15. Available online: https://www2.deloitte.com/us/en/insights/topics/marketing-and-sales-operations/global-marketing-trends/2020/purpose-driven-companies.html (accessed on 11 December 2021).

O'Connor, Nat. 2017. Three connections between rising economic inequality and the rise of populism. *Irish Studies in International Affairs* 28: 29–43. [CrossRef]

O'Neil, Cathy. 2016. *Weapons of Math Destruction: How Big Data Increases Inequality and Threatens Democracy*. New York: Crown Publishing Group.

O'Neill, Daniel W., Andrew L. Fanning, William F. Lamb, and Julia K. Steinberger. 2018. A good life for all within planetary boundaries. *Nature Sustainability* 1: 88–95. [CrossRef]

Oishi, Shigehiro, and Selin Kesebir. 2016. Income Inequality Explains Why Economic Growth Does Not Always Translate to an Increase in Happiness. Available online: https://www.academia.edu/15425048/Income_Inequality_Explains_Why_Economic_Growth_Does_not_Always_Translate_to_an_Increase_in_Happiness (accessed on 1 December 2021).

Organisation for Economic Co-operation and Development (OECD). 2017. Environmental Fiscal Reform: Progress, prospects, and pitfalls. Available online: http://www.oecd.org/tax/tax-policy/environmental-fiscal-reform-progress-prospects-and-pitfalls.htm (accessed on 12 December 2021).

Organisation for Economic Development and Co-operation (OECD). 2018. Poor Children in Rich Countries: Why We Need Policy Action. Available online: https://www.oecd.org/els/family/Poor-children-in-rich-countries-Policy-brief-2018.pdf (accessed on 12 December 2021).

Organisation for Economic Development and Co-operation (OECD). 2021. How's Life? Available online: https://www.oecdbetterlifeindex.org/ (accessed on 21 December 2021).

Organisation for Economic Development and Co-operation (OECD). 2022a. Tax and the Environment. Available online: https://www.oecd.org/ctp/tax-and-environment.htm (accessed on 8 July 2022).

Organisation for Economic Development and Co-operation (OECD). 2022b. Inequality. Available online: https://www.oecd.org/social/inequality.htm (accessed on 1 June 2022).

Ortiz-Ospina, Esteban, and Max Roser. 2020. Loneliness and Social Connections. Available online: https://ourworldindata.org/social-connections-and-loneliness (accessed on 12 November 2021).

Ostrom, Elinor. 1990. *Governing the Commons: The Evolution of Institutions for Collective Action*. New York: Cambridge University Press.

Ostry, Jonathan D., Prakash Loungani, and Davide Furceri. 2016. Neoliberalism: Oversold? *Finance & Development*. Available online: https://www.imf.org/external/pubs/ft/fandd/2016/06/pdf/ostry.pdf (accessed on 20 November 2021).

Otero, Gerardo, Gabriela Pechlaner, Giselle Liberman, and Efe Gürcan. 2015. The neoliberal diet and inequality in the United States. *Social Science & Medicine* 142: 47–55. [CrossRef]

Otto, Ilona M., Jonathan F. Donges, Roger Cremades, Avit Bhowmik, Richard J. Hewitt, Wolfgang Lucht, Johan Rockström, Franziska Allerberger, Mark McCaffrey, Sylvanus S. P. Doe, and et al. 2020. Social tipping dynamics for stabilizing Earth's climate by 2050. *PNAS* 117: 2354–65. [CrossRef] [PubMed]

Our World in Data. 2021. Self-Reported Life Satisfaction vs. GDP per Capita. Available online: https://ourworldindata.org/grapher/gdp-vs-happiness?xScale=linear (accessed on 30 November 2021).

Our World in Data. 2022. Economic Growth. Available online: https://ourworldindata.org/economic-growth (accessed on 12 January 2022).

Oxfam International. 2020. World's Billionaires Have More Wealth Than 4.6 Billion People. Available online: https://www.oxfam.org/en/press-releases/worlds-billionaires-have-more-wealth-46-billion-people (accessed on 12 November 2021).

Parker, Kim, and Juliana Menasce Horowitz. 2022. Majority of workers who quit a job in 2021 cite low pay, no opportunities for advancement, feeling disrespected. *Pew Research Center*. March 9. Available online: https://www.pewresearch.org/fact-tank/2022/03/09/majority-of-workers-who-quit-a-job-in-2021-cite-low-pay-no-opportunities-for-advancement-feeling-disrespected/ (accessed on 8 January 2022).

Parry, Ian W. H., Simon Black, and Nate Vernon. 2021. *Still not getting energy prices right: A global and country update of fossil fuel subsidies.* International Monetary Fund, Working Paper No. 2021/236. Available online: https://www.imf.org/en/Publications/WP/Issues/2021/09/23/Still-Not-Getting-Energy-Prices-Right-A-Global-and-Country-Update-of-Fossil-Fuel-Subsidies-466004 (accessed on 6 January 2022).

Pasqualino, Roberto, Aled W. Jones, Irene Monasterolo, and Alexander Phillips. 2015. Understanding global systems today—A calibration of the World3-03 model between 1995 and 2012. *Sustainability* 7: 9864–89. [CrossRef]

Passell, Peter, Marc Roberts, and Leonard Ross. 1972. The Limits to Growth. *The New York Times.* April 2. Available online: https://www.nytimes.com/1972/04/02/archives/the-limits-to-growth-a-report-for-the-club-of-romes-project-on-the.html (accessed on 12 October 2021).

People Staff. 1977. Lily Tomlin. *People.* December 26. Available online: https://people.com/archive/lily-tomlin-vol-8-no-26/ (accessed on 2 January 2022).

Pfeffer, Fabian T., and Alexandra Killewald. 2018. Generations of Advantage: Multigenerational Correlations in Family Wealth. *Social Forces* 96: 1411–42. [CrossRef] [PubMed]

Piketty, Thomas. 2014. *Capital in the Twenty-First Century.* Cambridge: Harvard University Press.

Piketty, Thomas. 2022. *A brief History of Equality.* Cambridge: Harvard University Press.

Pilling, David. 2018. *The Growth Delusion: Wealth, Poverty, and the Well-Being of Nations.* New York: Crown Publishing Group.

Pincetl, Stephanie, and Terri S. Hogue. 2015. California's new normal? Recurring drought: Addressing winners and losers. *Local Environment* 20: 850–54. [CrossRef]

Pink, Daniel H. 2011. *Drive: The Surprising Truth About What Motivates Us.* New York: Penguin Group.

Pizzol, Damiano, Carlo Foresta, Andrea Garolla, Jacopo Demurtas, Mike Trott, Alessandro Bertoldo, and Lee Smith. 2021. Pollutants and sperm quality: A systematic review and meta-analysis. *Environmental Science and Pollution Research* 28: 4095–103. [CrossRef]

Plenty of Gloom. 1997. *The Economist.* December 18. Available online: https://www.economist.com/christmas-specials/1997/12/18/plenty-of-gloom (accessed on 12 October 2021).

Polanyi, Karl. 2001. *The Great Transformation: The Political and Economic Origins of Our Time.* Boston: Beacon Press.

Pollan, Michael. 2009. *In Defense of Food: An Eater's Manifesto.* New York: Penguin Random House.

Polman, Paul, and Andrew S. Winston. 2021. *Net Positive: How Courageous Companies Thrive by Giving More Than They Take.* Brighton: Harvard Business Review Press.

Porges, Stephen W. 2017. Vagal pathways: Portals to compassion. In *The Oxford Handbook of Compassion Science*. Edited by Emma Seppala, Emiliana Simon-Thomas, Stephanie L. Brown, Monica C. Worline, C. Daryl Cameron and James R. Doty. New York: Oxford University Press, pp. 189–202.

Preston, Stephanie D. 2022. *The Altruistic Urge: Why We're Driven to Help Others*. New York: Columbia University Press.

Price, Carter C., and Kathryn A. Edwards Edwards. 2020. *Trends in Income From 1975 to 2018*. Santa Monica: RAND Corporation. Available online: https://www.rand.org/pubs/working_papers/WRA516-1.html (accessed on 12 October 2021).

PricewaterhouseCoopers. 2021. What's Next for America's Workforce Post-COVID-19? Available online: https://www.pwc.com/us/en/services/consulting/business-transformation/library/workforce-pulse-survey.html?_hsenc=p2ANqtz-9f4f3a_LrbSclTcHaaCoN4pn2jdOiG7ntx1SNVmf_e0FRd_L8s4hhtDr_tnCZqfP0kzvv9 (accessed on 6 December 2021).

Project Implicit. 2022. Project Implicit: Take a Test. Available online: https://implicit.harvard.edu/implicit/selectatest.html (accessed on 12 February 2022).

Quinn, Robert E., and Anjan V. Thakor. 2019. *The Economics of Higher Purpose: Eight Counterintuitive Steps for Creating a Purpose-Driven Organization*. Oakland: Berrett-Koehler Publishers.

Rainie, Lee, Scott Keeter, and Andrew Perrin. 2019. Trust and Distrust in America. *Pew Research Center*. July 22. Available online: https://www.pewresearch.org/politics/2019/07/22/trust-and-distrust-in-america/ (accessed on 12 October 2021).

RAND. 2022. Delphi Method. Available online: https://www.rand.org/topics/delphi-method.html (accessed on 11 February 2022).

Randers, Jorgen. 2000. From limits to growth to sustainable development or SD (sustainable development) in a SD (system dynamics) perspective. *System Dynamics Review* 16: 213–24. [CrossRef]

Randers, Jorgen, Johan Rockström, Per Espen Stoknes, Ulrich Golüke, David Collste, and Sarah Cornell. 2018. *Transformation Is Feasible—How to Achieve the Sustainable Development Goals within Planetary Boundaries*. Stockholm: Stockholm Resilience Centre. Available online: https://www.stockholmresilience.org/download/18.51d83659166367a9a16353/1539675518425/Report_Achieving%20the%20Sustainable%20Development%20Goals_WEB.pdf (accessed on 5 October 2021).

Rapid Transition Alliance. 2018. Buen Vivir: The Rights of Nature in Bolivia and Ecuador. Available online: https://www.businessinsider.com/corporate-communism-is-killing-us-2009-10 (accessed on 1 October 2021).

Ratigan, Dylan. 2009. Corporate Communism Is Killing Us. *Business Insider*. October 7. Available online: https://www.businessinsider.com/corporate-communism-is-killing-us-2009-10 (accessed on 5 October 2021).

Raworth, Kate. 2017. *Doughnut Economics: 7 Ways to Think Like a 21st Century Economist*. White River Junction: Chelsea Green Publishing Co.

Rebalance Earth. 2022. Funding Biodiversity to Combat Climate Change. Available online: https://www.rebalance.earth/ (accessed on 9 February 2022).

Reddy, Sujana, Vamsi Reddy, and Sandeep Sharma. 2018. *Physiology, Circadian Rhythm.* Treasure Island: StatPearls Publishing.

Refinitif. 2022. Environmental, Social and Corporate Governance—ESG. Available online: https://www.refinitiv.com/en/financial-data/company-data/esg-data (accessed on 2 February 2022).

Reinhart, Carmen, and Kenneth Rogoff. 2009. *This Time Is Different: Eight Centuries of Financial Folly.* Princeton: Princeton University Press.

Renda, Andrea, Sylvia Schwaag Serger, Daria Tataj, Andrew Morlet, Darja Isaksson, Francisca Martins, Montserrat Mir Roca, César Hidalgo, Ailin Huang, Sandrine Dixson-Declève, and et al. 2022. *Industry 5.0, a Transformative Vision for Europe: Governing Systemic Transformations towards a Sustainable Industry.* Brussels: European Commission. Available online: https://data.europa.eu/doi/10.2777/17322 (accessed on 9 April 2022).

Robertson, Brian J. 2015. *Holacracy: The New Management System for a Rapidly Changing World.* New York: Henry Holt and Company.

Robeyns, Ingrid. 2019. What, if anything, is wrong with extreme wealth? *Journal of Human Development and Capabilities* 20: 251–66. [CrossRef]

Roca, Thomas. 2011. Subjective Well-Being: Easterlin Paradox, the (Decreasing) Return(s)? From Log to Square, New Evidence from Wealthier Data. *SSRN Electronic Journal.* [CrossRef]

Roemer, John E. 2000. *Equality of Opportunity.* Cambridge: Harvard University Press.

Roncoroni, Alan, Stefano Battiston, Marco D'Errico, Grzegorz Hałaj, and Christoffer Kok. 2019. *Interconnected Banks and Systemically Important Exposures.* Working Paper Series, No. 2331. Frankfurt: European Central Bank. Available online: https://www.ecb.europa.eu/pub/pdf/scpwps/ecb.wp2331~{}ab59126ee2.en.pdf (accessed on 11 November 2021).

Roopsind, Anand, Brent Sohngen, and Jodi Brandt. 2019. Evidence that a national REDD+ program reduces tree cover loss and carbon emissions in a high forest cover, low deforestation country. *Proceedings of the National Academy of Sciences* 116: 24492–99. [CrossRef]

Rosenberg, Marshall. 2012. *Living Nonviolent Communication: Practical Tools to Connect and Communication Skillfully in Every Situation.* Louisville: Sounds True Inc.

Roser, Max. 2014. Fertility Rate. Available online: https://ourworldindata.org/fertility-rate (accessed on 9 October 2021).

Routledge, Clay, and Taylor A. FioRito. 2021. Why Meaning in Life Matters for Societal Flourishing. *Frontiers in Psychology* 11: 601899. [CrossRef] [PubMed]

Rözer, Jesper Jelle, and Beate Volker. 2016. Does income inequality have lasting effects on health and trust? *Social Science and Medicine* 149: 37–45. [CrossRef] [PubMed]

Rubio-Licht, Nat, Veronica Irwin, and Protocol Team. 2022. Amazon's Union Fight: Here's What's Happening Now. *Protocol.* February 8. Available online: https://www.protocol.com/amazon-union (accessed on 9 June 2022).

Rufrancos, Hector Gutierrez, Madeleine Power, Kate E. Pickett, and Richard Wilkinson. 2013. Income inequality and crime: A review and explanation of the time-series evidence. *Sociology and Criminology* 1: 103. [CrossRef]

Ruiz, Daniel, Marisol Becerra, Jyotsna S. Jagai, Kerry Ard, and Robert M. Sargis. 2018. Disparities in environmental exposures to endocrine-disrupting chemicals and diabetes risk in vulnerable populations. *Diabetes Care* 41: 193–205. [CrossRef] [PubMed]

Saeed, Khalid. 2014. Policy Space and System Dynamics Modeling of Environmental Agendas. Available online: http://digitalcommons.wpi.edu/ssps-papers/4 (accessed on 9 October 2021).

Saez, Emmanuel, and Gabriel Zucman. 2019. *The Triumph of Injustice: How the Rich Dodge Taxes and How to Make Them Pay.* New York: Norton & Company.

Salesforce. 2022. The Path to Integrating ESG Data into the Core of a Business—Behind Salesforce and Accenture's Expanded Partnership. Available online: https://www.salesforce.com/news/stories/sustainability-faq-davos-2021-update/ (accessed on 2 February 2022).

Sassen, Saskia. 2014. *Expulsions: Brutality and Complexity in the Global Economy.* Cambridge: The Belknap Press of Harvard University Press.

Schaeffer, Katherine. 2020. 6 Facts about Economic Inequality in the U.S. *Pew Research Center.* February 7. Available online: https://www.pewresearch.org/fact-tank/2020/02/07/6-facts-about-economic-inequality-in-the-u-s/ (accessed on 2 November 2021).

Schneider Electric. 2021. Digital economy and climate impact. Available online: https://www.se.com/ww/en/insights/electricity-4-0/digital-for-recovery/digital-economy-and-climate-impact.jsp (accessed on 1 June 2022).

Schopenhauer, Arthur. 2004. *The Wisdom of Life.* Mineola: Dover Publications.

Schwab, Klaus. 2021. *Stakeholder Capitalism: A Global Economy that Works for Progress, People and Planet.* Hoboken: Wiley.

Schwartz, Tony, Jean Gomes, and Catherine McCarthy. 2011. *Be Excellent at Anything: The Four Keys to Transforming the Way We Work and Live.* New York: Simon and Schuste.

Securities and Exchange Commission. 2022. SEC Response to Climate and ESG Risks and Opportunities. Available online: https://www.sec.gov/sec-response-climate-and-esg-risks-and-opportunities (accessed on 2 February 2022).

Senge, Peter. 1994. *The Fifth Discipline: The Art and Practice of the Learning Organization.* New York: Currency Doubleday.

Sharp, Leith. 2017. *Sustainability Leadership in the 21st Century.* Cambridge: Harvard Extension School—Sustainability Program.

Simmons, Matthew R. 2000. *Revisiting the Limits to Growth: Could the Club of Rome Have Been Correct, After All?* Eugene: Mud City Press. Available online: http://www.mudcitypress.com/PDF/clubofrome.pdf (accessed on 12 October 2021).

Singh, Laxmi, Madhu Anand, Saroj Singh, and Ajay Taneja. 2020. Environmental toxic metals in placenta and their effects on preterm delivery-current opinion. *Drug and Chemical Toxicology* 43: 531–38. [CrossRef] [PubMed]

Slaughter, Anne-Marie. 2017. *The Chessboard and the Web: Strategies of Connection in a Networked World*. New Haven: Yale University Press.

Smillie, Susan. 2017. From Sea to Plate: How Plastic Got into Our Fish. *The Guardian*. February 14. Available online: https://www.theguardian.com/lifeandstyle/2017/feb/14/sea-to-plate-plastic-got-into-fish (accessed on 22 November 2021).

Smith, Brendan L. 2012. The case against spanking. *Monitor on Psychology* 43: 60. Available online: http://www.apa.org/monitor/2012/04/spanking (accessed on 2 February 2022).

Solow, Robert M. 1973. Is the end of the world at hand? *Challenge* 16: 39–50. [CrossRef]

Solow, Robert M. 2003. Dumb and Dumber in Macroeconomics. Available online: https://pdfcoffee.com/solow-dumb-and-dumber-in-macroeconomicspdf-pdf-free.html (accessed on 22 December 2021).

Solow, Robert. 2008. The State of Macroeconomics. *The Journal of Economic Perspectives* 22: 243–46.

Sorkin, Andrew Ross, Jason Karaian, Sarah Kessler, Stephen Gandel, Michael J. de la Merced, Lauren Hirsch, and Ephrat Livni. 2022. Larry Fink Defends Stakeholder Capitalism. *New York Times*. January 18. Available online: https://www.nytimes.com/2022/01/18/business/dealbook/fink-blackrock-woke.html#:~{}:text=As%20the%20founder%20and%20C.E.O.,that%20his%20investment%20firm%20manages.&text=He%20again%20stressed%20that%20E.S.G.,factor%E2%80%9D%20in%20its%20investment%20assessments (accessed on 22 December 2021).

Standing Bear, Luther. 2006. *Land of the Spotted Eagle*. Lincoln: University of Nebraska Press.

Steffen, Will, Katherine Richardson, Johan Rockström, Sarah E. Cornell, Ingo Fetzer, Elena M. Bennett, Reinette Biggss, Tephen R. Carpenter, Wim De Vries, Cynthia A. De Wit, and et al. 2015. Planetary boundaries: Guiding human development on a changing planet. *Science* 347: 1259855. [CrossRef] [PubMed]

Stein, Jeremy C. 2010. Securitization, Shadow Banking, and Financial Fragility. *Daedalus* 139: 41–51. [CrossRef]

Sterman, John D. 1994. Learning in and about complex systems. *System Dynamics Review* 10: 291–330. [CrossRef]

Sterman, John D. 2000. *Business Dynamics: Systems Thinking and Modeling for a Complex World*. Boston: Irwin/McGraw-Hill.

Stiglitz, Joseph. 2012. *The price of inequality: How today's divided society endangers our future*. New York: W.W. Norton.

Stiglitz, Joseph E., Amartya Sen, and Jean-Paul Fitoussi. 2010. *Mismeasuring Our Lives: Why GDP Doesn't Add Up*. New York: The New Press.

Stiglitz, Joseph, Jean-Paul Fitoussi, and Martine Durand. 2019. *Measuring What Counts: The Global Movement for Well-Being*. New York: The New Press.

Surowiecki, James. 2005. *The Wisdom of Crowds*. New York: Doubleday; Anchor.

Sustainable Development Report. 2021. *Rankings*. Available online: https://dashboards.sdgindex.org/rankings (accessed on 1 February 2022).

Sverdrup, Harald, and K. Vala Ragnarsdóttir. 2014. Natural resources in a planetary perspective. *Geochemical Perspectives* 3: 129–341. [CrossRef]

Sverdrup, Harald, and K. Vala Ragnarsdóttir. 2015. 40 Years after Limits to Growth: The World3 System Dynamics Model and Its Impacts. [PowerPoint Slides]. Available online: http://www.wrforum.org/wp-content/uploads/2015/09/Limits-to-growth.pdf (accessed on 2 October 2021).

Swiss Re. 2020. A Fifth of Countries Worldwide at Risk from Ecosystem Collapse as Biodiversity Declines, Reveals Pioneering Swiss Re Index. Available online: https://www.swissre.com/media/news-releases/nr-20200923-biodiversity-and-ecosystems-services.html (accessed on 22 February 2022).

Tàbara, J. David, Niki Frantzeskaki, Katharina Hölscher, Simona Pedde, Kasper Kok, Francesco Lamperti, Jens H. Christensen, Jill Jäger, and Pam Berry. 2018. Positive tipping points in a rapidly warming world. *Current Opinion in Environmental Sustainability* 31: 120–29. [CrossRef]

Tajitsu, Naomi. 2019. Toyota to Give Royalty-Free Access to Hybrid-Vehicle Patents. *Reuters*. April 3. Available online: https://www.reuters.com/article/us-toyota-patents/toyota-to-give-royalty-free-access-to-hybrid-vehicle-patents-idUSKCN1RE2KC (accessed on 12 February 2022).

Taskforce on Climate-related Financial Disclosures (TCFD). 2017. *Final Report: Recommendations of the Task Force on Climate-Related Financial Disclosures.* Available online: https://www.fsb-tcfd.org/publications/final-recommendations-report (accessed on 2 December 2021).

Tax Justice Network. 2022. The State of Tax Justice 2020. Available online: https://taxjustice.net/reports/the-state-of-tax-justice-2020/ (accessed on 3 February 2022).

Tax Policy Center. 2020. Historical Highest Marginal Income Tax Rates. Available online: https://www.taxpolicycenter.org/statistics/historical-highest-marginal-income-tax-rates (accessed on 2 October 2021).

Thaler, Richard H. 2000. From Homo Economicus to Homo Sapiens. *The Journal of Economic Perspectives* 14: 133–41. [CrossRef]

Thaler, Richard H., and Cass R. Sunstein. 2009. *Nudge: Improving decisions about health, wealth, and happiness.* New Haven: Yale University Press.

The Editorial Board. 2020. Virus Lays Bare the Frailty of the Social Contract. *Financial Times.* April 3. Available online: https://www.ft.com/content/7eff769a-74dd-11ea-95fe-fcd274e920ca (accessed on 13 December 2021).

The University of Texas at Austin. 2019. New Oil and Gas Production Technologies. Available online: https://www.strausscenter.org/energy-and-security/new-oil-and-gas-production-technologies.html (accessed on 3 December 2021).

Thibodeaux, Wanda. 2017. Why Working in 90-Minute Intervals Is Powerful for Your Body and Job, According to Science. *Inc.* January 27. Available online: https://www.inc.com/wanda-thibodeaux/why-working-in-90-minute-intervals-is-powerful-for-your-body-and-job-according-t.html (accessed on 5 February 2022).

Tollefson, Jeff. 2020. Why deforestation and extinctions make pandemics more likely. *Nature* 584: 175–76. [CrossRef] [PubMed]

Tong Lee, Sheryl Tian. 2022. China Unveils ESG Reporting Program in Bid to Catch Global Peers. *Bloomberg*. May 31. Available online: https://www.bloomberg.com/news/articles/2022-05-31/china-unveils-esg-reporting-program-in-bid-to-catch-global-peers (accessed on 3 June 2022).

Trang, Brittany, Yuli Li, Xiao-Song Xue, Mohamed Ateia, K. N. Houk, and William R. Dichtel. 2022. Low-temperature mineralization of perfluorocarboxylic acids. *Science* 377: 839–45. [CrossRef] [PubMed]

Trasande, Leonardo. 2019. *Sicker, Fatter, Poorer: The Urgent Threat of Hormone-Disrupting Chemicals to Our Health and Future . . . and What We Can Do About It.* Boston: Mariner Books.

Trebeck, Katherine, and Jeremy Williams. 2019. *The Economics of Arrival: Ideas for a Grown-up Economy.* Bristol: Bristol University Press.

Triodos Bank. 2022. Our Mission and Vision. Available online: https://www.triodos.com/about-us (accessed on 30 January 2022).

Troller-Renfree, Sonya V., Molly A. Costanzo, Greg J. Duncan, Katherine Magnuson, Lisa A. Gennetian, Hirokazu Yoshikawa, Sarah Halpern-Meekin, Nathan A. Fox, and Kimberly G. Noble. 2022. The impact of a poverty reduction intervention on infant brain activity. *Proceedings of the National Academy of Sciences of the United States of America* 119: e2115649119. [CrossRef] [PubMed]

Turban, Stephen, Laura Freeman, and Ben Waber. 2017. A study used sensors to show that men and women are treated differently at work. *Harvard Business Review*. October 23. Available online: https://hbr.org/2017/10/a-study-used-sensors-to-show-that-men-and-women-are-treated-differently-at-work (accessed on 6 December 2021).

Turner, Graham M. 2008. A comparison of the Limits to Growth with 30 years of reality. *Global Environmental Change* 18: 397–411. [CrossRef]

Turner, Graham M. 2012. On the cusp of global collapse? Updated comparison of the Limits to Growth with historical data. *GAIA* 21: 116–124. [CrossRef]

Turner, Graham M. 2013. The limits to growth model is more than a mathematical exercise: Reaction to R. Castro. 2012. Arguments on the imminence of global collapse are premature when based on simulation models. GAIA 21/4: 271–273. *GAIA-Ecological Perspectives for Science and Society* 22: 18–19. [CrossRef]

Turner, Graham. 2014. *Is Global Collapse Imminent?* Melbourne: Melbourne Sustainable Society Institute, The University of Melbourne. Available online: http://sustainable.unimelb.edu.au/sites/default/files/docs/MSSI-ResearchPaper-4_Turner_2014.pdf (accessed on 3 October 2021).

Tyson, Ruth. 2020. *An Open Letter to the Union of Concerned Scientists: On Black Death, Black Silencing, and Black Fugitivity: An Affirmation of Black Life from a Concerned Black Human.* [Google Docs]. Available online: https://docs.google.com/document/d/132Ow3_FYcTQdc73pyAe0V_IIoIzq4nktYpZ-V4MZRTY/edit (accessed on 30 November 2021).

Union of Concerned Scientists (UCS). 2022. Founding Document: 1968 MIT Faculty Statement. Available online: https://www.ucsusa.org/about/history/founding-document-1968-mit-faculty-statement (accessed on 30 November 2021).

United Nations (UN). 1992. *Report of the United Nations Conference on Environment and Development.* Available online: https://www.un.org/en/development/desa/population/migration/generalassembly/docs/globalcompact/A_CONF.151_26_Vol.I_Declaration.pdf (accessed on 18 December 2021).

United Nations (UN). 2021a. 'The Adolescence of Humanity is Coming to an End', Declares UK Prime Minister. *UN News.* September 22. Available online: https://news.un.org/en/story/2021/09/1100882 (accessed on 5 December 2021).

United Nations (UN). 2021b. 'Tremendously Off Track' to meet 2030 SDGs: UN chief. *UN News.* July 12. Available online: https://news.un.org/en/story/2021/07/1095722 (accessed on 7 December 2021).

United Nations Conference on Trade and Development (UNCTAD). 2013. Alternative scenarios for the world economy. In *Trade and Development Report 2013: Adjusting to the Changing Dynamics of the World Economy.* New York: United Nations, pp. 37–46. [CrossRef]

United Nations Department of Economic, Social Affairs, and Statistical Division (UN DESA Statistical Division). 2021a. International Standard Industrial Classification of All Economic Activities (ISIC), Rev. 4. Available online: https://unstats.un.org/unsd/publication/seriesm/seriesm_4rev4e.pdf (accessed on 15 December 2021).

United Nations Department of Economic and Social Affairs, Statistical Division (UN DESA Statistical Division). 2021b. System of National Accounts 1993–1993 SNA. Available online: https://unstats.un.org/unsd/nationalaccount/sna1993.asp (accessed on 15 December 2021).

United Nations Development Programme (UNDP). 2021a. Education Index. Available online: http://hdr.undp.org/en/indicators/103706 (accessed on 16 December 2021).

United Nations Development Programme (UNDP). 2021b. Sources of Data Used. Available online: http://hdr.undp.org/en/statistics/understanding/sources (accessed on 16 December 2021).

United Nations Development Programme (UNDP). 2021c. Human Development Index (HDI). Available online: http://hdr.undp.org/en/indicators/137506 (accessed on 16 December 2021).

United Nations Development Programme (UNDP). 2021d. Did the HDI Rankings Change for Many Countries in 2019? Available online: http://hdr.undp.org/en/content/did-hdi-rankings-change-many-countries-2019 (accessed on 16 December 2021).

United Nations Environment Programme (UNEP). 2002. Global Environment Outlook 3. Available online: https://www.unenvironment.org/resources/global-environment-outlook-3 (accessed on 2 November 2021).

United Nations Environment Programme (UNEP). 2016. Half the World to Face Severe Water Stress by 2030 Unless Water Use Is "Decoupled" from Economic Growth, Says International Resource Panel. Available online: https://www.unep.org/news-and-stories/press-release/half-world-face-severe-water-stress-2030-unless-water-use-decoupled (accessed on 6 January 2022).

United Nations Environment Programme (UNEP). 2021a. Food Waste Index Report 2021. Available online: https://www.unep.org/resources/report/unep-food-waste-index-report-2021 (accessed on 24 January 2022).

United Nations Environment Programme (UNEP). 2021b. Adapt to Survive: Business Transformation in a Time of Uncertainty. Available online: https://www.unep.org/resources/publication/adapt-survive-business-transformation-time-uncertainty (accessed on 6 June 2022).

United Nations Environment Programme Financial Initiative (UNEP Financial Initiative). 2019. Fiduciary Duty in the 21st Century Final Report. Available online: https://www.unepfi.org/publications/investment-publications/fiduciary-duty-in-the-21st-century-final-report/ (accessed on 24 May 2022).

United Nations Water (UN Water). 2021. UN World Water Development Report 2021. Available online: https://www.unwater.org/publications/un-world-water-development-report-2021/#:~{}:text=The%202021%20edition%20of%20the,valuing%20water%20infrastructure%20for%20water (accessed on 4 February 2021).

United States Census Bureau (USCB). 2021a. Income and Poverty in the United States: 2020. Available online: https://www.census.gov/library/publications/2021/demo/p60-273.html (accessed on 3 December 2021).

United States Census Bureau (USCB). 2021b. Historical Income Tables: People. Available online: https://www.census.gov/data/tables/time-series/demo/income-poverty/historical-income-people.html (accessed on 3 December 2021).

United States Census Bureau (USCB). 2021c. Income, Poverty and Health Insurance Coverage in the United States: 2020. Available online: https://www.census.gov/newsroom/press-releases/2021/income-poverty-health-insurance-coverage.html (accessed on 3 December 2021).

United States Department of Agriculture. 2020. In 2019, 13.6 Percent of Households with Children Were Affected by Food Insecurity. Available online: https://www.ers.usda.gov/data-products/chart-gallery/gallery/chart-detail/?chartId=58390 (accessed on 24 January 2022).

United States Geological Survey (USGS). 2021a. Historical Statistics for Mineral and Material Commodities in the United States. Available online: https://www.usgs.gov/centers/nmic/historical-statistics-mineral-and-material-commodities-united-states (accessed on 23 November 2021).

United States Geological Survey (USGS). 2021b. Commodity Statistics and Information. Available online: https://www.usgs.gov/centers/nmic/commodity-statistics-and-information (accessed on 23 November 2021).

Urbina, Dante A., and Alberto Ruiz-Villaverde. 2019. A Critical Review of Homo Economicus from Five Approaches. *American Journal of Economics and Sociology* 78: 63–93. [CrossRef]

US Department of Housing and Urban Development. 2021. 2020 Annual Homeless Assessment Report Part 1 to Congress. Available online: https://www.huduser.gov/portal/datasets/ahar/2020-ahar-part-1-pit-estimates-of-homelessness-in-the-us.html (accessed on 3 December 2021).

US Environmental Protection Agency. 2002. Wastewater Technology Fact Sheet. Available online: https://www3.epa.gov/npdes/pubs/living_machine.pdf (accessed on 1 July 2022).

van Daalen, Kim Robin, Sarah Savić Kallesøe, Fiona Davey, Sara Dada, Laura Jung, Lucy Singh, Rita Issa, Christina Alma Emilian, Isla Kuhn, Ines Keygnaert, and et al. 2022. Extreme events and gender-based violence: A mixed-methods systematic review. *The Lancet: Planetary Health* 6: 504–23. [CrossRef]

van den Bergh, Jeroen C. J. M. 2011. Environment versus growth—A criticism of "degrowth" and a plea for "a-growth". *Ecological Economics* 70: 881–90. [CrossRef]

van Sebille, Erik, Chris Wilcox, Laurent Lebreton, Nikolai Maximenko, Britta Denise Hardesty, Jan A. van Franeker, Marcus Eriksen, David Siegel, Francois Galgani, and Kara Lavender Law. 2015. A global inventory of small floating plastic debris. *Environmental Research Letters* 10: 124006. [CrossRef]

Vargish, Thomas. 1980. Why the person sitting next to you hates limits to growth. *Technological Forecasting & Social Change* 16: 179–89. [CrossRef]

Vensim. 2022. Free Downloads. Available online: https://vensim.com/free-download/ (accessed on 3 January 2022).

Vermeulen, P. J., and D. C. J. de Jongh. 1976. 'Dynamics of growth in a finite world'—Comprehensive sensitivity analysis. *IFAC Proceedings Volumes* 9: 133–45. [CrossRef]

Vidal, John. 2011. Bolivia enshrines natural world's rights with equal status for Mother Earth. *The Guardian*. April 10. Available online: https://www.theguardian.com/environment/2011/apr/10/bolivia-enshrines-natural-worlds-rights#:~{}:text=They%20include%3A%20the%20right%20to,structure%20modified%20or%20genetically%20altered (accessed on 29 September 2021).

Wachowski, Lana, Lilly Wachowski, and Joel Silver. 2003. *The Matrix Revolutions. [Motion picture]*. Burbank: Warner Bros.

Wackernagel, Mathis. 2021. Thanks for LTG piece. *Email*.

Walker, Matthew. 2017. *Why We Sleep: Unlocking the Power of Sleep and Dreams*. New York: Simon & Schuster.

Weir, Kirsten. 2017. Why we believe alternative facts. *Monitor on Psychology* 48: 5. Available online: http://www.apa.org/monitor/2017/05/alternative-facts (accessed on 12 November 2021).

Wellbeing Economy Alliance (WEAll). 2022a. WEAll Is a Collaboration of Organisations, Alliances, Movements and Individuals Working towards a Wellbeing Economy, Delivering Human and Ecological Wellbeing. Available online: https://weall.org/ (accessed on 31 January 2022).

Wellbeing Economy Alliance (WEAll). 2022b. Wellbeing Economy Governments. Available online: https://weall.org/wego (accessed on 31 January 2022).

Wellbeing Economy Alliance (WEAll). 2022c. Wellbeing Economy Policy Design Guide. Available online: https://wellbeingeconomy.org/wp-content/uploads/Wellbeing-Economy-Policy-Design-Guide_Mar17_FINAL.pdf (accessed on 31 January 2022).

Westley, Frances, Per Olsson, Carl Folke, Thomas Homer-Dixon, Harrie Vredenburg, Derk Loorbach, John Thompson, Måns Nilsson, Eric Lambin, Jan Sendzimir, and et al. 2011. Tipping Toward Sustainability: Emerging Pathways of Transformation. *Ambio* 40: 762. [CrossRef] [PubMed]

Wike, Richard, Janell Fetterolf, Shannon Schumacher, and J. J. Moncus. 2021. Citizens in advanced economies want significant changes to their political systems. *Pew Research Center*. October 21. Available online: https://www.pewresearch.org/global/2021/10/21/citizens-in-advanced-economies-want-significant-changes-to-their-political-systems/ (accessed on 31 January 2022).

Winfrey, Oprah. 2012. President Obama to Oprah: 'I'm not done yet'. *Huffpost*. October 19. Available online: https://www.huffpost.com/entry/president-obama-barack-obama-oprah_n_2041826 (accessed on 2 January 2022).

Woody, Todd. 2013. New Drilling Technologies Could Give Us So Much Oil, the Climate Won't Stand a Chance. *Quartz*. August 21. Available online: https://qz.com/117504/new-drilling-technologies-could-give-us-so-much-oil-the-climate-wont-stand-a-chance/ (accessed on 4 January 2022).

World Bank (WB). 2012. Natural Capital: Managing Resources for Sustainable Growth. In *Inclusive Green Growth: The Pathway to Sustainable Development*. pp. 105–31. Available online: http://siteresources.worldbank.org/EXTSDNET/Resources/Inclusive-Green-Growth-Chapter5.pdf (accessed on 31 December 2021).

World Bank (WB). 2018. Decline of Global Extreme Poverty Continues but Has Slowed: World Bank. Available online: https://www.worldbank.org/en/news/press-release/2018/09/19/decline-of-global-extreme-poverty-continues-but-has-slowed-world-bank (accessed on 31 October 2021).

World Bank (WB). 2020. *Poverty and Shared Prosperity 2020: Reversals of Fortune*. Washington, DC: World Bank. [CrossRef]

World Bank (WB). 2021a. *The Changing Wealth of Nations 2021: Managing Assets for the Future*. Washington, DC: World Bank. Available online: https://openknowledge.worldbank.org/handle/10986/36400 (accessed on 31 January 2022).

World Bank (WB). 2021b. Population, Total. Available online: https://data.worldbank.org/indicator/SP.POP.TOTL (accessed on 13 February 2022).

World Bank (WB). 2021c. Fertility Rate, Total (Births per Woman). Available online: https://data.worldbank.org/indicator/SP.DYN.TFRT.IN (accessed on 13 February 2022).

World Bank (WB). 2021d. Death Rate, Crude (per 1000 People). Available online: https://data.worldbank.org/indicator/SP.DYN.CDRT.IN (accessed on 13 February 2022).

World Bank (WB). 2021e. Manufacturing, Value Added (Constant 2015 US$). Available online: https://data.worldbank.org/indicator/NV.IND.MANF.KD (accessed on 3 December 2021).

World Bank (WB). 2021f. Gross Capital Formation (Constant 2015 US$). Available online: https://data.worldbank.org/indicator/NE.GDI.TOTL.KD (accessed on 3 December 2021).

World Bank (WB). 2021g. Government Expenditure on Health, Total (% of GDP). Available online: https://data.worldbank.org/indicator/SH.XPD.CHEX.GD.ZS (accessed on 7 December 2021).

World Bank (WB). 2021h. [Data files]. Unpublished Excel Files. Retrievable from World Bank at Request.

World Bank (WB). 2021i. Life expectancy at birth, total (years)—United States. Available online: https://data.worldbank.org/indicator/SP.DYN.LE00.IN?end=2020&locations=US&start=2010 (accessed on 1 December 2021).

World Bank (WB). 2022a. *Global Economic Prospects*. Washington, DC: World Bank. [CrossRef]

World Bank (WB). 2022b. Globally, Countries Lose $160 Trillion in Wealth Due to Earnings Gaps Between Women and Men. Available online: https://www.worldbank.org/en/news/press-release/2018/05/30/globally-countries-lose-160-trillion-in-wealth-due-to-earnings-gaps-between-women-and-men (accessed on 7 February 2022).

World Bank (WB). 2022c. The World Bank in Bhutan. Available online: https://www.worldbank.org/en/country/bhutan/overview (accessed on 1 August 2022).

World Economic Forum (WEF). 2017. Shaping the Future of Global Food Systems: A Scenarios Analysis. Available online: http://www3.weforum.org/docs/IP/2016/NVA/WEF_FSA_FutureofGlobalFoodSystems.pdf (accessed on 2 February 2022).

World Economic Forum (WEF). 2018. Executive Summary. Available online: http://reports.weforum.org/global-risks-2018/executive-summary (accessed on 1 February 2022).

World Economic Forum (WEF). 2022a. The New Narratives Lab. Available online: https://www.weforum.org/new-narratives-lab/programme (accessed on 1 February 2022).

World Economic Forum (WEF). 2022b. Stakeholder Capitalism: Over 70 Companies Implement the ESG Reporting Metrics. Available online: https://www.weforum.org/impact/stakeholder-capitalism-50-companies-adopt-esg-reporting-metrics/?utm_source=sfmc&utm_medium=email&utm_campaign=2777903_Am22-AgendaDaily-25May2022&utm_term=&emailType=Agenda%20Weekly (accessed on 1 February 2022).

World Health Organization (WHO). 2017. National Chemicals Registers and Inventories: Benefits and Approaches to Development. Available online: https://www.euro.who.int/__data/assets/pdf_file/0018/361701/9789289052948-eng.pdf (accessed on 11 February 2022).

World Health Organization (WHO). 2021. Global Health Expenditure Database. Available online: http://apps.who.int/nha/database (accessed on 7 December 2021).

Wright, Stephanie L., and Frank J. Kelly. 2017a. Threat to human health from environmental plastics. *BMJ* 358: j4334. [CrossRef] [PubMed]

Wright, Stephanie L., and Frank J. Kelly. 2017b. Plastic and human health: A micro issue? *Environmental Science and Technology* 51: 6634–47. [CrossRef] [PubMed]

Zakrzewski, Anna, Kedra Newsom Reeves, Michael Kahlich, Maximilian Klein, Andrea Real Mattar, and Stephan Knobel. 2020. Managing the Next Decade of Women's Wealth. *BCG*. April 9. Available online: https://www.bcg.com/publications/2020/managing-next-decade-women-wealth (accessed on 4 February 2022).

Zheng, Guomao, Erika Schreder, Jennifer C. Dempsey, Nancy Uding, Valerie Chu, Gabriel Andres, Sheela Sathyanarayana, and Amina Salamova. 2021. Per- and Polyfluoroalkyl Substances (PFAS) in Breast Milk: Concerning Trends for Current-Use PFAS. *Environmental Science & Technology* 55: 7510–20. [CrossRef]

MDPI
St. Alban-Anlage 66
4052 Basel
Switzerland
Tel. +41 61 683 77 34
Fax +41 61 302 89 18
www.mdpi.com

MDPI Books Editorial Office
E-mail: books@mdpi.com
www.mdpi.com/books